Julia M. Wright

Botany

The Story of Plant Life

Julia M. Wright

Botany
The Story of Plant Life

ISBN/EAN: 9783337119683

Printed in Europe, USA, Canada, Australia, Japan

Cover: Foto ©berggeist007 / pixelio.de

More available books at **www.hansebooks.com**

BOTANY

The Story of
Plant Life

BY JULIA Mac NAIR WRIGHT

Author of " Nature Readers, Seaside and Wayside,"
" Astronomy," etc.

ILLUSTRATED

" Flower in the crannied wall !
   I pluck you out of the crannies ;
I hold you here, root and all in my hand,
Little flower—if I could but understand
What you are, root and all, and all in all,
   I should know what God and man is."
        —*Tennyson.*

PHILADELPHIA

THE PENN PUBLISHING COMPANY

1898

CONTENTS

CONTENTS

BOTANY

—

CHAPTER I

THE STORY OF THE ROOT

JANUARY

"Never quite shall disappear
The glory of the circling year."

THE EXTENT OF PLANT LIFE

PLANT life extends over almost the entire surface of
the globe. If, as is highly probable, the waters of
the Polar seas, under the perpetual ice-cap, have
their algæ, or sea-weeds, then the only places really
destitute of vegetation are the crests of the highest
mountains, the highest regions of glaciers and ever-
enduring snows.

The seas are full of sea-weeds, from the most minute to the largest known plants. Under the snow
of the Arctic region are beds of lichens, while upon

5

the surface are the tiny algæ, known as "crimson snow."

The faces of great rocks bear films of lichens, which, when observed under the microscope, appear as groves and forests, wherein play minute animals, invisible to unaided eyes. The long-dead logs and rails in fences are clothed with fairy gardens, knobs of gold, cups of scarlet, frost-work of gray, white, black, green.

Parasites and epiphytes hang high upon the branches of the trees; gray-green mistletoe, flaming orchids; while about the grasses and little shrubs twines the dodder, also a parasite; and from the spreading roots of pines lift the wax-white beech-drops.

The waters of ponds are covered with what people loosely call "green scum from stagnation," but which is nothing more nor less than a floating bed of strange plants. Water-lilies and other aquatic growths spring from the mud-beds of ponds, and along the margins of rivers. The stones under water afford root-hold for plants.

Plant life invades the store-closet and pantry; it takes hold upon loaves of bread, the tops of jars of preserves, pickles, and jellies — "mold" we call it. If we examine it carefully, this "mold" turns out to be a crowded collection of little plants, beautifully shaped and highly colored. Were it not for this invasion of our domestic privacy by plant life, we should have no yeast and no vinegar.

Deep down into the wells climb the lichens and algæ; the marl-pit and the lime-kiln are no sooner abandoned by men than nature prepares to send her plant-hosts to take possession. If there is a crack in a rock, a seed falls into it, germinates, rives the rock; lo! life triumphs over death, and action over inaction. If the greatest cities in the world were now abandoned of men, within fifty years the houses would be draped with lichens, vines, fungi; the streets would be a tangle of weeds, creepers, briars; great forest trees would rise on every side; the battalions of the birds, the cohorts of the winds would bring the seeds and superintend the reconverting of the abodes of humanity to tangled wilderness.

Vegetable life preceded man upon the earth, and keeps equal pace with him in his progress upon it. Wherever man abides the plants accompany him. We think of the frigid zones as denuded of vegetation; even there, in the brief summers, poppies bloom and grasses wave close to the retreating snow-line. So the Alpine eidelweiss and gentian climb the heights and nod under the edges of the glaciers.

The existence of plant life is a condition upon which human existence depends. The presence of plants is as needful to the life of man as air, light, and water.

THE UTILITY OF PLANT LIFE

Why is the plant needful to the life of animals? For food: Plant life must have preceded animal life, because the original food of the animal is the plant. Plants are the great food-making shops. Only plants can take mineral substances from soil or water and convert them into food stuff. Animals cannot feed upon minerals; the minerals must be converted into vegetable material before animals can assimilate them. The first food of the first animal, whatever that animal was—some sea-creature of simple form, probably—was some small and simple sea-weed, which had existed before the animal.

The plant can live without the animal; the animal cannot live without the plant. The plant can take its food directly from soil, air, and water; the animal must have the plant for its middleman in its dealings with nature. There are, it is true, carnivorous animals, creatures feeding upon flesh, either living, dead or decayed. This flesh, however, is of creatures that are vegetable eaters. The lion devours the antelope, kid or giraffe. These are eaters of herbs. We humans drink milk and eat meat, as well as vegetable substances, but our animal food

and drink come to us from vegetable-eaters. We
use fish largely; these feed immediately or remotely
upon vegetables.

An ample proportion of our medicines, oint-
ments, and other healing agents come to us from
plants.

To the plant world we look for a large amount of
our clothes, as linen and cotton directly; silk from
the mulberry-fed silk-worm; wool from the grass-
eating sheep; leather from vegetable-devouring
mammals.

The plant world affords us fuel, immediately as
wood, or remotely from the coal beds, which once
were forests.

Our houses and our furniture are largely of
wood, contributed by the plant world. Our horses
and oxen serve us with strength supplied by plant-
food. The presence of vegetation increases that
downfall of rain which fills our springs, wells,
streams, cisterns.

The plants, with their million million busy
mouths, devour from the air seeds of pestilence,
converting them into beauty and utility; they pour
into our atmosphere oxygen, and sip out of it
noxious carbonic acid gas. Thus on every hand, for
the luxuries and the necessities of our lives, we are
debtors to the plant.

THE STUDY OF PLANT LIFE

This beneficent and abounding vegetation has always attracted the observation and aroused the curiosity of the human race. From the earliest times plants have been more or less profoundly investigated, their beauty and utility alike provoking attention. Amid mistakes, myths, superstitions, the inquiry has progressed, each generation arriving at some real knowledge to add to the stock gained before. Errors have been corrected, discoveries made, mysteries unsealed, the study becoming more and more beautiful under the search-lights of science. In these days a young student may easily know as much about the plant world as Solomon, Esculapius or Pliny.

No pursuit is more conducive to health, happiness, refinement of character than the study of plants. This pleasing pursuit keeps one largely in the open air, in company with pure and helpful thinking; the subjects for study are spread liberally on every hand. Without books or instruments, simply by spending leisure time and careful observation upon the plants about his feet, one can become a fairly good botanist. Aided by books, teachers, a cabinet, microscope, and collections, one can make rapid advances in plant lore. The passion for botanical work grows by indulgence. For all these reasons parents and teachers should lead the minds of the young to the interests of plant life.

The study of botany is not "merely the pulling of plants to pieces to find out certain dry names"; it is directed to every department of the plant world, and invades and brings back trophies from the wonder and romance land of the vegetable kingdom.

One may begin with fossil plants, may continue with the algæ of the waters, the lichens on the walls, the mosses in the wood, or find ample occupation in the life mysteries of the splendid hosts of the flowering plants.

THE BORDER-LAND

One may begin to study the most highly specialized blossoms, and insensibly be led, step by step, down to the lower planes of life. The rose beckons the student back along the line of plant ascent, and beguiles to an investigation of that lesser relation, the strawberry, and more humble yet, the star-faced cinquefoil. Lower still, the microscope piloting the way, we may be among the invisible, so far as the naked eye is concerned. We study, bent over the pond, glass in hand, the duck-weed; and here is something lashing the water, showing motion; is it an animal or a vegetable? We have reached the border-land where these two meet—and mingle. A few decades ago we were told that the sponges marked this indefinite land. Now, the sponges have their indubitable place in the animal kingdom, and the microscope must direct our way to the wonderful " What is it?" animal or vegetable?

Cuvier found four characteristics which duly marked the animal from the plant: 1. A food reservoir or stomach; 2. Circulatory system; 3. Locomotion and sensibility, provided for by a more complicated body containing nitrogen; 4. Respiration. In the years that have gone since Cuvier, we have found that some animals lack the digestive apparatus and some plants possess it. The test mark of the circulatory system was practically given up by Cuvier himself; chemistry has annihilated the third distinction, for nitrogen is as essential to plants as to animals; finally, the respiration of plants is as fully marked as in many species of animals.

What is the test-mark then, the broad arrow of the animal kingdom? Who knows? Huxley tells us that Professor Tyndall asked him to examine some microscopic objects traveling in water "by spasmodic wiggles."

" What do you call them?" asked Tyndall.

" They may be animals, and then again they may be vegetables," said Huxley.

Mr. Tyndall replied that "he would as soon be-
lieve a sheep to be a plant."

And yet, Huxley says that, after long study, he
remains where he was at first : " There is no reason
why this minute monad may not be an animal, and
there is equally no reason why it may not be a
vegetable."

Such is that border-land between the animal and
the plant, where the foot of the diligent botanist
may ultimately tread. Into such Dark Continent
we do not propose to go. In this simple little work
a guide-book to the beginning of the way, we merely
commence the study of Plant Life with the earth and
end in the earth.

FROM ROOT TO FRUIT

There is no season of the year in which we cannot
pursue the story of plant life. Every month offers its
theme. Let us devote January to making the ac-
quaintance of the root. When a seed germinates it
sends forth a sprout, placed perpendicularly. The
upper growing point is the stem or ascending axis.
It has affinity for light and heat ; it produces
buds, which develop into leaves, blossoms, and
other parts of a plant. This stem also produces
roots, for its downward portion has affinity for darkness and moisture; it is an earth-lover, produces no

buds, and becomes the root or roots of the new plant.

It is no matter of mere chance which axis of the new sprout goes up, and which down ; the stem will make a complete bend around the parent seed, if that is needed to reach the surface of the earth ; while its proper lower part will be at equal trouble to bury itself in the soil.

It is a popular fallacy that stems grow from roots. On the contrary roots grow from stems. Even in the perennial roots of plants which die down to the ground each autumn, the root does not send up fresh stems, a portion of true stem remains under ground on the root crown, and this it is which sends up bud-bearing stems. The distinguishing mark of the true root is that it bears no buds, no leaves, no scales.

There are underground stems and there are also aerial roots. We have all noted the fringe of roots starting from the lower joints of a cornstalk, and growing perhaps several inches before entering the earth. The heavy stalk needs these additional moorings, as tents need their cords. Some trees, as the banyan, put forth roots from branches high in air. These must grow many feet before they can enter the earth. None of these roots put forth leaves. We find on brook sides sometimes great portions of apparent roots, laid bare by the falling away of covering earth.

In a few years some of these put on a heavy bark, like that of the trunk, and throw out tufts of leaves or twigs. This proves them to be branches sent forth underground, at the base of the stem.

A potato is a thick underground stem. The true roots of the potato-vine are the string-like fibres which grow from the lower portion of the erect stem. What we call the potato is a stem thickened for a food reservoir. The eyes on this potato are buds; by each one is a thin oval or triangular slip of scale—a modified, abnormal leaf. The potato-eyes being planted, with a portion of potato for food, begin to grow and send out stems and leaves.

What is a lily bulb—a root? Cut it in two, behold, it is a stem with two buds for next year at its base. These thick white scales are closely packed bleached leaves or leaf-bases, which never develop further, but which are detailed to furnish food to be absorbed by the growing stem. A case nearly similar is afforded by the onion.

Take a sweet potato; that has not the eyes nor the scales that an Irish potato shows, to prove its status as a stem. What is it? A sweet potato is a thickened portion of the true root. Most of the roots of a sweet potato vine are fibrous, but in places they swell out, forming solid masses, which are store-houses of sugar and starch for plant-food. The tend-

ency of the sweet potato in a wild state to form
these food masses has been emphasized by careful
cultivation.

Roots are called fibrous when composed entirely
of string-like clusters, as, for instance, a bunch of
grass or corn roots.

Fleshy roots are those that thicken and expand,
as in the beet, turnip, and carrot. From these
thickened parts true roots start forth. Look at a
carrot; the rind or skin of the upper portion of this
beautiful golden wedge has, by exposure at the surface of the ground, become tough, fibrous, greenish;
it has put forth a cluster of green leaves. The portion deeper in the earth has a thin, porous skin,
bright golden color, and sends forth root fibres. This
carrot is not, then, pure root throughout. The upper
part of the wedge is stem. Let us carefully lay open
the thick top, and cut out one by one the leaf stems.
We can now see how they take their rise, and distin-
guish the true stem character of the upper part of
the carrot. The meaning of the long thick wedge is
again storehouse—food reservoir. The tendency of
the carrot to lay up food in this way has been
immensely increased by cultivation. Examine a
wild carrot, or parsnip, and in the very much smaller
" root wedge " we shall detect the true stem character. A primary root is the first root which starts

from the young stem. It is also called the main root, or tap root, when it goes on growing without branching, except close beneath the stem. From the primary root may start out hairs, which develop into rootlets. When a number of nearly equal roots start out at once from the base of the stem they are called " multiple primary roots." There are more hairs near the tips of roots than anywhere else upon them. These hairs are the chief root channels for gathering nourishment.

All the surface of the root absorbs moisture, and this moisture holds in solution mineral foods, which are converted into vegetable substance by the plant. Every plant is a wonderful laboratory.

At the tip of each root is a hardened, scale-protected point, to enable it to work its way into the ground, just as the toes of moles and like digging animals are protected by nails. The root tip has also a sucker, for drawing up moisture. This is the chief mouth of the root. Although the pores of the entire root-surface absorb freely, the ends of the rootlets are the chief feeders.

A fibrous root is a feeding root. Fleshy roots are feeding roots through their surfaces and rootlets; they are also storehouses, or pantries, for laying up food for the future use of the plant. This stored-up food is largely starch and sugar.

2

The office of the root is: I. To obtain nutriment for the plant. II. To hold or moor the plant to the ground. III. To store up food for future use.

Most roots are fixed in the earth. There are also air or aerial roots, such as the ivy and trumpet-creeper send forth from the sides of the stem to serve as hands, to grapple the supports upon which they climb. Other aerial roots penetrate the bark of branches upon which they have fixed themselves, and then suck out sap for food.

Roots are called annual, when they live but one year; biennial, when they live for two years; perennial, or ever-enduring, when they live for many years. Some plants live over a thousand years. Ivies are noted of five hundred years old; a grape-vine eight hundred years old; trees a thousand years old.

Roots naturally perennial may become annual when transplanted to colder than their native places. Cold, however, does not kill all roots, it merely suspends much of their activity.

When we walk about in January and see the earth frozen, or covered with snow, we need not fancy that all is still and dead under ground. There are millions of mouths below the surface, taking their rest, and feeding but little; there are other millions of plant store-houses, food-full, for coming summer;

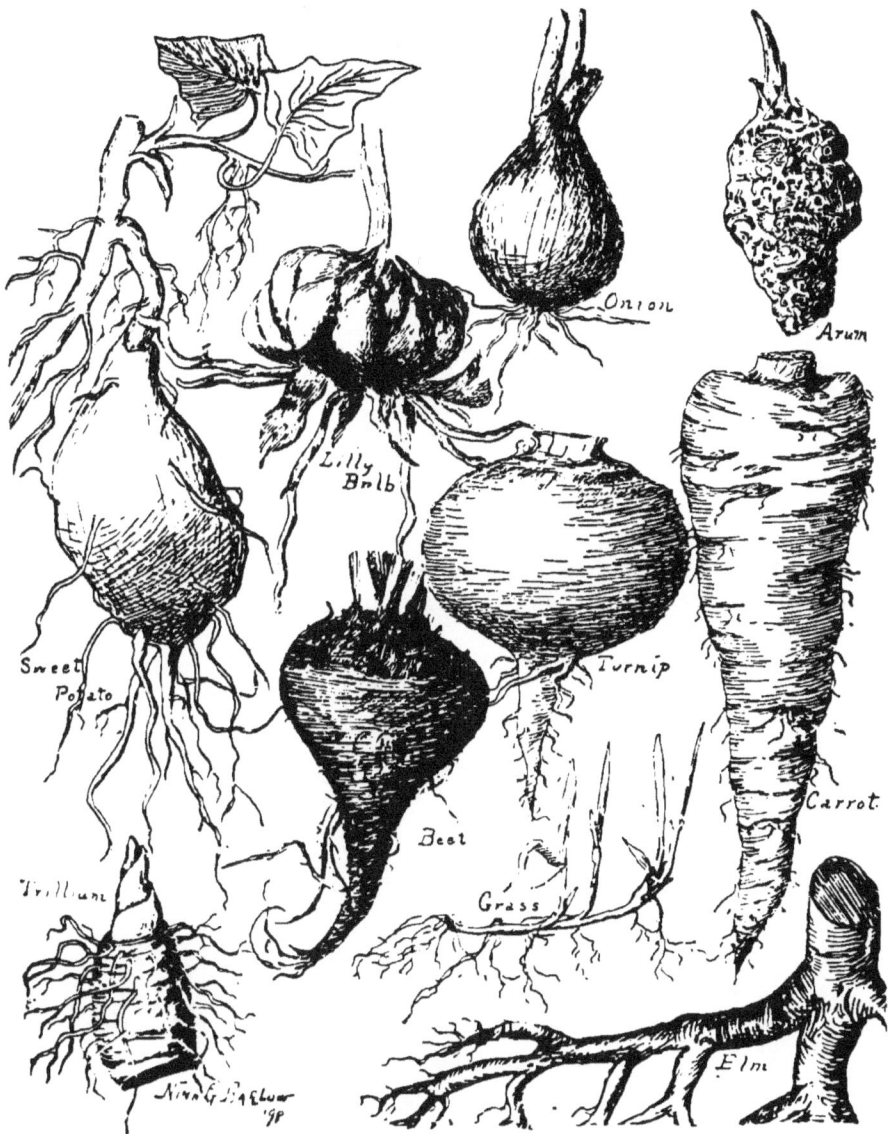

Onion

Arum

Lilly
Bulb

Sweet
Potato

Turnip

Carrot

Beet

Grass

Trillium

Elm

TREASURES OF DARKNESS

there are countless underground stems; and all
are waiting for the call of spring, to renew their activity, and clothe the earth with beauty.

In this cold, forbidding month of January the
plant does not refuse to unveil to us some of the romance, the mystery, and the economy of its life.
Pull up from some plat of unfrozen earth a few living roots and hold them against the light, after giving them a slight shake. Tiny particles of earth are
now seen about the lower part of the root, not ad-
hering to the epidermis, but held, perhaps, a line
away, as if the root were covered with a brown-
dotted lace veil. Examine closely, with the microscope if possible. Each atom of earth is held by a
minute hair. These hairs are of great importance
in the economy of the root. They adhere so very
closely to the soil that they absorb from it the very
slightest trace of moisture, if it be no more than
such a light film as would be found if one breathed
against glass. That earth must be dry as lava on
Vesuvius at midday from which these hairs could
not extract some particles of moisture. In times of
great drought these fine hairs allow the plant to
gather enough fluid to enable them to survive, and
when the drooping plant is watered the hairs most
speedily gather up and distribute the precious drops.
These hairs are tubes, but they are tubes closed at

the ends, and they do not absorb by capillary attraction, but by their entire surface, as their walls are porous. When we speak of tubes, pores, surface, in this connection, and pause to consider that in many cases each entire hair is almost microscopic in its smallness, we wonder, as we are always wondering when we study plant life, at the singular perfection in detail, at the marvels of minutiæ, at intricacy and accuracy in smallness. The life of each of these tiny root hairs is short, as the root grows and its thimble-cap, its armored apex advances, new hairs develop on its surface, and the previous ones die. Always there is a fresh series of absorbing hairs, these not only sucking moisture from the soil, but each one is a little laboratory, where chemical processes are carried on. The wall of the hair-cell divides the liquid which it receives into its different parts. It separates the crystals from the albumen or starch, and so passes it on, nicely prepared food-stuff for the plant.

Hairs are not found merely on roots. They occur on every part of the plant. In this month of January, so unsympathetic with plant life in our northern zone, we can make interesting studies in plant hairs of various kinds. In a wide botanical sense any appendages of the plant skin are hairs.

Here is a rose bush, raising its dry, reddish stems

against a wall; here, along the garden path, are our rows of raspberry and blackberry vines, all well provided with prickles, which we carelessly call " thorns." Real thorns, however, are woody, will not come off with the skin, and are, in fact, hardened, undeveloped branches. The prickles strip off with the epidermis, and are really included properly under the general term of hairs.

These prickles keep cows, sheep, and other grazing animals from eating up the bushes, which have a sweetish, aromatic taste, and are sufficiently tender to afford very tempting morsels to browsing beasts in winter, hungering for something fresh, as we humans long for the first crisp salads, radishes, and other spring vegetables. Were it not for these blessed prickles, not a rose nor a berry would greet our eyes next summer, but nature, having thoughtfully detailed this army of sharp pikesmen to defend her sleeping children, they will awaken in the spring in health and beauty.

Wandering a little farther afield in search of vegetable hairs, we see a tall, rough, dead rod, set with seed vessels at the top—the dried stalk of a mullein. At its base we shall find a large rosette of greenish-gray, thick leaves; some dry and dead, after a summer's expansion; some young and still succulent, having unfolded late in the autumn. All these leaves are covered so thickly with hairs that they resemble leaves cut out of coarse flannel or felt. Every part of the mullein plant, except the petals, stamens, and pistils is hairy. It is a very Esau among plants; nothing but a donkey would graze upon it. Exactly why a plant of so little value and attractiveness as the mullein should be so carefully defended, we cannot see; the fairer sisters of its kin are left almost naked to cold winds and nipping teeth.

The hairs on the mullein plant may serve to make it unpalatable, but their primary use is probably as clothing for preserving an equal temperature and shielding the plant from cold and wet. Very many Alpine and far-northern plants are so protected. Another instance of hairs as clothing is afforded us this January day if we examine the tree buds. Some of them, it is true, seem to be very carelessly clad. They are of the hardier varieties. Some buds are waxed or varnished, but very many are hair-clad, fur-wrapped, nature thus dressing her precious babies for their winter outing.

Many seed pods are lined with a delicate network of hairs, which are especially numerous and closely-woven at the suture or seam of the pod, where it will open when the seeds are ripe.

Here in the fields we find some lingering pods of

the silk-weed, which takes its name from the great
mass of fine silky hairs which fills its seed vessels.
Nature provided in this fashion for the dissemination of the seed. We loosen a handful of the dainty
hairs from the pod, hold them to the air, and away
they sail! They remind us of so many other wind-
blown seeds! In spring the dandelion clocks are
like fairy balloons tilting along the top of the
grasses. The first breese loosens them; away they
go, silver boats with silken sails, voyaging to fairy
land, to bring back to the children next spring another
golden age! Many a farmer who endures dandelions for old love's sake has cursed the equally-well-
provided thistle, and wished dire wreck to overtake
every one of its little boats sailing the seas of air;
but thistles, like some other creations, seem to thrive
particularly well on curses! Prominent among all
seeds provided with hairy sails stands one that has a
world-wide and political importance—King Cotton.
What would the races of men do if that vast mass
of snowy, elastic, tough hairs which enfold the oily
seed of the cotton plant met disaster? Ruin, naked-
ness, starvation would shriek around a world, all
for want of certain vegetable hairs! Other plant-
hairs have a less noble reputation. Here is a withered root of the crimson clover, an introduced
variety. The Agricultural Department of the

United States sends out word that the hairs, plentiful on the calyx of this beautiful trefoil, are often dangerous to cattle, as they form balls in the intestines.

If we close our January botanical ramble in a greenhouse, we are still met by fresh examples of vegetables hairs. Here is the stem of an elephant ear, hirsute as the face of a Russian sledge driver; and here is a palm, with stem wrapped and twisted with long interlaced hairs until it looks like a ravelled door mat. Such hairs secure an equalized temperature, keep out wet when the plants grow in their native state in lands with rainy seasons, and serve in dry seasons to hinder the evaporation of the moisture of the plant. This study of plant hairs will come up again.

CHAPTER II

THE STORY OF THE STEM

FEBRUARY

" Ancient pines !
Ye bear no record of the years of man.
Spring is your sole historian."

THE stem is that part of the plant which bears all
the other organs. It produces roots from its base;
leaves, branches, flowers, fruit from its sides and
apex. As the human body maintains the head and
limbs, sheltering within it the various organs of
breathing, nutrition, circulation, etc., the stem serves
the plant as a body, upbearing and supplying the
organs.

One department of the plant world is truly stemless. These stemless plants have neither roots, leaves,
nor seeds. They are mere expansions of cellular
tissue, beautiful, useful, wonderful, varied. At present
a few of their names, only, are mentioned with some
of the most common examples. They form too difficult a study for the amateur in botany.

Some plants are apparently, but not truly, stemless. The leaves and flowers seem to spring directly

25

from the roots. If we uncover the base of the leaf cluster we find a thickened root-like portion called "the crown of the root," which is really a shortened stem, sending leaves and bloom from its upper surface, and roots from the lower surface, The dandelion, primrose, and cowslip are examples.

Take a cyclamen plant, lay it bare of earth, and you find a fleshy button, from the top of which grow leaves, from the bottom roots. This button is a stem, which, as a bright child said, "grows sideways, not upwards." Jack-in-the-pulpit, beloved of all children, has a solid bulb, called a corm, or very short stem. The roots are sent out, fringe-like, in a ring about the thickest part, above them rise leaves and Jack. The cyclamen button-stem enlarges and lasts for years; the corm of Jack-in-the-pulpit forms the first year, is eaten by Jack for his health the next year, another corm forming during this process, just above the roots,

All this storing up of fleshy stem is to provide food for the upper parts of the plant. The lily, onion, hyacinth form bulbs of thick, bleached imperfect leaves, about the base of the stem, as described in the previous chapter; these are devoured in the process of seed making.

In the woods in May, you find a plant with a slender stem, gracefully bent by the weight of leaves and flowers. The leaves grow on the upper side of the stem, the little white bell-blossoms droop in the arch beneath them. This plant is the "Solomon's Seal." Later in the season these flowers have given place to blueberries containing seeds. Digging up this plant you will find that the slender stem is but a branch from a thick underground stem, from six to ten inches long; this lies horizontally, having at one end a bud preparing for a leaf and flower stem for next year. The underground stem is marked at regular intervals by a round, smooth scar, indicating where a stem died down to the undergound portion the previous year. Each year the new bud develops fresh roots, as it sends up a fresh stem. A length, as above given, having been made, each year a joint and scar die off at one end of the underground root, while the new bud forms at the other, decreasing at one end about as fast as it increases at the opposite extremity. The scars on the Solomon's Seal underground stem mark each year of plant life; they keep the record of its growth.

Another spring wild-flower that has an odd stem is the Trillium. It looks like a lily, red, yellow or white, and it receives its name from the three-fold division of its parts. If you dig up the short, thick underground stem, you will see that it grows in rings, as if one ring were laid upon another. From each ring start fibrous roots. The lower end of the stem looks as if it had been bitten short off by some animal. Every ring on this buried stem marks a year of the plant's life; thus nature keeps some of her chronicles. We will mention but one other underground stem, that of the potato.

The potato has three kinds of stems. First, green stems which bear leaves; second, green stems which bear flowers and seeds only; third, underground stems which are the fleshy tubers that we eat. The chief part of the food stuff of the potato plant is carried down the leaf stem into these thickened underground branches, and there stored up, chiefly in the form of starch. The great supply of a very nourishing food stuff which the potato is able to lay up, has made it a valuable article of diet for man. The care of the farmer has been directed to increasing this tendency of the potato to store up food. The potato comes from the uplands of the Andes of Chili, and since its discovery, in 1563, it has been cultivated in almost every country cool enough for its growth.

A remarkable form of stem is shown in the cactus family. These plants grow in hot, sandy countries, subject to long droughts. The amount of food and moisture which they are enabled to lay up in their stems fits them well for their habitation. These cactus stems are green and leafless; the thick rind acts the part of leaves in preparing food. They shoot up into tall, strong columns, expand into melon-shaped masses, or into that succession of fleshy, rudely-oval discs with which we are familiar in the "pear cactus;" also there are other fantastic forms. The stem bears the sessile, usually showy blossoms, and is abundantly furnished with sharp prickles, to prevent the plant's destruction by thirsty animals.

"What do stems bear?" Let us ask this question and answer it as we ramble. "Leaves, flowers, roots," we speak briskly; "fruit," more slowly. "Nothing else?" "No." They bear much more than these. Excrescences of the rind sharpen into prickles, as in the blackberry and rose, protecting the plant from devastation. The honey-locust, hawthorn, thorn-apple, and plum have large spines which are stunted, hardened branches, or stem tips. The barberry has thorns formed of undeveloped leaves, hardened almost like steel. We know that these are altered leaves, as they grow each in an axil, under a bud. These are all protective appendages to preserve the plant for seed bearing.

Stems also produce tendrils for climbing purposes. Some tendrils grow out straight, until they reach a place of support; there they adhere by means of a flat disc at the tendril tip. As soon as they are fast the tendril shrinks or shortens, drawing the vine to the place of support. The Virginia creeper exhibits this wonderful contrivance. Some stems, as the ivies, send out rootlets to fasten upon the rough surface of walls and to secure support. The grape vine has branching tendrils, which are really altered, suppressed branches, converted to climbing appliances. An examination of tendrils will afford delightful study. One variety of tendrils reaches out after a stick or stem, and then wraps and curls about it, ring after ring, with wonderful precision.

Plants are divided into herbs, bushes, shrubs, trees, according as the stem is soft, short, annual, or large, woody, and enduring. The herb has a soft, usually green stem, never very tall, dying to the ground each year, as the marigold, lily, etc. A bush has a woody stem of moderate height, and a rind thicker than that of an herb; it is perennial, as the rose, spirea, currant. A shrub has a still thicker, stronger, taller, more woody, rougher-skinned stem; it is also perennial, as the lilac, snowball, syringa. It may be from ten to twenty feet high, and is thickly branched. A tree is much larger in every way than the shrub. It has numerous great roots, a wide spread of branches, a strong bark, and lasts from thirty or forty to a thousand years.

Again, stems are divided by their manner of growth; there are straight, upright, much-branching trunks, as the pine, oak, elm, and others; there are climbing and twining stems, as the grape and morning glory; stems that run along the ground, as the pumpkin; stems that sprawl lazily in every direction. Some stems are hollow, as reeds, grasses, oats, and wheat; some stems are bare runners, which root at their tips, producing new plants. The strawberry is of this last variety, and it is a pleasing experiment to set a strawberry plant in a well-tended plat of ground and see how many new plants it will produce from its runners in the course of two or more years. The runner having rooted at the tip, dies off and leaves the new plant to send out fresh runners.

Plants which gardeners tell you " multiply by the root," are really growing in this fashion, from runners under ground. Many stems can be laid down and covered with earth, and thus forced to send out roots and buds at their joints or nodes, so producing new plants.

Whenever you find underground a long shoot sending out rootlets and scales, you may know by the scales that this is a stem, for scales are aborted leaves, and only stems produce leaves. Thus the potato has at each of its buds or eyes a thin scale, which is an undeveloped leaf.

In all woody stems, the growth of the year is not only in height but in girth; next the bark there is produced a new ring of material each year. Examine a log that has been cut smoothly across; you will find that from the centre to the bark it is a series of rings. These rings are of unequal width. The narrow rings represent dry seasons, where the tree had not enough moisture to grow rapidly; the wide rings speak of seasons when the conditions of growth were better. Thus the tree writes the history of the seasons through which it has passed.

The young stem lifted into the air differs very markedly from the root, in that the tip of the root is covered with a protective, horny cap, while the growing point of the stem has no such cap. The apex of the stem is a bud; this bud is more or less closely wrapped in scales, which seem to be covered with a thin varnish to protect the bud from too much moisture. The growth of the stem presses this bud upward, the scales unfold, the bud opens, leaves and branches are formed. New leaves are sent forth at the axils, but the growing point, the bud at the apex, is still the top of the plant.

If this apex is cut off it is not replaced by a new one, but the vitality of the stem deprived of this point of activity, throws out fresh branches lower down.

The gardener frequently cuts away the growing point of plants to force them to make more leaves and branches, or to send out a richer bloom. For instance, if the tips of chrysanthemum stalks are clipped off, the bloom thrown out from the sides of the stems will be much finer and more profuse.

Every plant begins with a single cell. This cell has an outside case, a soft, jelly-like mass, and within that an atom of fluid, called cell-sap. We are not now speaking of things that can be seen by the naked eye. A good microscope is needed to make cells visible. The jelly-like part is the most important; some cells dispense with the sac. A cell enlarges, divides, forms new cells; so plant structure is built up.

Cells united form cellular tissue; this is most abundant in thick, fleshy plants. There are millions of cells in a very small plant; countless billions in a great tree. Cells built together, in various ways, form the tall thick stems, the divided branches, the many-shaped leaves, the variously beautiful flowers which to our minds compose the plant.

As the cells lie close together, sometimes the walls give way, and from a long line of cells a tube is formed. Some cells are built together in such fashion as to leave spaces between which form tubes also. When we consider that the most delicate silken hair

3

on a child's head is a hollow tube, we can guess
something of the fineness of the tubes which, banded
together form stems.

Why should there be tubes? Why should you
have a throat to convey your food and drink to your
digestive organs? The roots, by their mouths and
general surface, absorb moisture from the earth. This
moisture holds, in a dissolved state, many minerals,
as salt, lime, soda, iron, chalk, silex, and others. The
moisture so gathered up, and laden, ascends through
the stem, passing up the tubes, and is distributed to
the leaves and general surface of the plant.

By the leaves, or the green rind serving instead of
leaves, this material is turned into plant stuff. It is
then sent down through the tubes which distribute
the material to all parts of the plant, to build up
more cells and tissues. The long, thickened, united
cells make up the woody fibre of trees, the fine tissues of the leaves.

We can think of the stem as a vast series of tubes
placed side by side, perpendicularly, their walls
more or less thick, forming different substances in
the plant for different uses. These fibres are so firm
that they build up the timber known as iron wood,
and other kinds of hard wood, as solid and heavy
almost as iron. Thread, ropes, cloth, are spun and
woven from these vegetable fibres; cloth is even

obtained by beating all the soft parts from the fibres, and pounding the fibres into a kind of felt.

If leaves, and some kinds of stems, are soaked in water until the soft parts are decayed, this decayed portion can be gently brushed from the woody frame-work, and the beautiful structure clearly exhibited.

Cells are not always placed in columns, or straight lines; they may be spiral, or in rings, or in six-sided canals, or in porous sacs laid one against another.

These interior tissues of the plant serve for circulation, for breathing, for the digesting of food, for building up of frame-work and of fleshy parts—as the organs of the body perform such functions for animals.

The bark and rinds of stems and branches are formed of tougher tissue, hardened by the air, sun, and storms; it is often spread with a thin varnish-like coating, to protect the inner portions from rain, and from insects. Some stems, as was noted in the previous chapter, are provided with hairs, or fine down; they look as if dressed in fur jackets.

Near this rind or bark, lie the sap vessels. If the bark is cut through all about the trunk, the tree will die for lack of circulation of sap. This cutting around of the bark near the base of the trunk is called girdling; it is a way taken to kill trees which are to be hewn down. A tree cut in this way early in spring, after the sap begins to ascend is so full of sap that it will put forth leaves, which will live some time on the sap which has already gone up into the branches.

February affords a very good season for the study of the stem. In the mild days we can search for underground stems; we have the trees and shrubs; also the house plants show us herb stems; the vegetables in the cellar give us further specimens of tubers, corms, bulbs. By the aid of a microscope very wonderful and beautiful structures will be discovered. On a sunny late February day we linger in some woodland walk; there is already the promise of spring in the mild air and the clear blue depths of sky; this brilliant atmosphere brings out very emphatically the marvelous tracery of stems and branches, making the denuded trees almost as beautiful objects as those in their leafy prime. Our admiration of these graceful and intricate divisions and multiplied subdivisions is not lessened when we realize that all this network of trunk, branch, limb, twig on each tree is the development of some one single original bud. In these days in February, trunks and branches lose that dry dull look which, for three months, has made them appear as dead trees; there is a certain freshness and warmth of color in bark and bud which suggests the circulation of the sap; something of that difference which appears between a dead and a sleeping human body. In fact, in February the sap has already begun to rise, as is testified in many parts of the country by the active work in the sugar maple groves, where gallons of sap are being collected and boiled. From almost any stem, if now cut across, water will drip. This water, or sap, the life-blood of the plant, is forced up by root pressure, which is especially great in the early months of the year. One botanist found that in a portion of grapevine this pressure of sap up from the root had a force sufficient to support a column of mercury over thirty-two inches high. In some plants, especially young and tender specimens, the root drives up the sap at such a furious rate that it presses it through the epidermis, and causes it to stand in drops on the surface. A very simple experiment proves that this upward rush of sap is caused by pressure from the root; take any stem from which drops are issuing, cut it from the main stock or root, and place it in water—no more drops will appear. Of course, these drops are not pure water, but hold various substances in solution, and these mineral particles will be built up into plant stuff.

As we examine stems in February or March and

notice the amount of sap issuing from any cut, we recall what seems an extraordinary fact—that in July and August stems being cut do not leak or bleed, and yet in those summer months that underground engine, the root, must be pumping up liquor faster than ever to maintain all that luxuriance of leafy growth. The explanation of this is simple : the plant, when well provided with leaves, is devouring for its own nutrition so much sap—" using it up " so fast—and is also losing so much moisture to the air, by reason of transpiration, or giving off of water in vapor, that no matter how hard the root works no water remains standing in the vessels to run out when cut. Here, of course, we except those peculiar, fleshy stems of some plants of arid countries, which stems are always a storehouse of water.

In our February walk, when the absence of leaves and of herbage gives good opportunity for the study of woody perennial stems, we at once notice how the shape of these is affected by the circumstances of their place of growth. In a thick wood the stems are slender and tall, shooting up in their effort to obtain air and sunlight. When valuable trees are to be cultivated, either as landscape ornaments or as timber supplies, one early process is to " thin them out," giving each one room to make a large stem or bole. Travelers in tropic lands tell us that in the crowded forests of hot countries trees like the palm and others having naturally an erect stem, have this stem often reduced to very small diameter and immensely elongated, so that they seem more like vines than trees, clinging to and rambling over their stouter neighbors. A tree very common in damp woods in New Jersey and other parts of the Northern States is the hornbeam, which has not a round, but a square stem, the angles being well defined, and this stem, unable to support its own weight, leans on adjacent trees, reaching from one to another, its crown emerging finally in some very unexpected place far from its roothold. A good example of a woody climbing stem in our forests is the wild grape, which, in a favorable locality, becomes very large, the main stem several inches in diameter, and by its weight finally breaking down some large tree upon which it has seized for support. If we examine these stems closely we shall find them covered with buds, still folded firmly, and we remember that that good old botanist, Linnæus, who seemed always to be in close sympathy with plants and trees, called these nicely-protected buds *hibernacula* or winter quarters of young branches.

If the blueness of a February sky and some sudden, unexpected softness of the air beguiles us to think of something like flowers, we must not expect to find them in the woods, but in the swamps; not on the hillsides, but down along the brooks. There, even in these last days of February, we may find a blossom. Even when films of ice and little rifts of snow lie about their roots, the willow trees, or rather shrubs—for these especial willows do not reach the dignity of trees—begin to bloom.

Who does not know and love these soft, silken "pussies," which give their name to the pussy willows. They are of a dull pale purple color, and one of the daintiest bouquets imaginable is a large bunch of the leafless willow stems, closely set with the soft purple pussies. In a day or two, if our willow branches are kept in water and placed where the sun can touch them, the hazy purple turns to a film of gold. Looking closely, we shall find that the crowded stamens have matured their pollen and are ready to toss it off at a touch in a golden shower. If we had left the willow stems by their brook some gentle breeze, or a brush from the wing of some passing bird, or some early bee, would have sent the yellow atoms flying.

Perhaps close by our willow shrubs some alders are growing; the alder stems are hung with scale-covered tassels called catkins; a few days of sunshine and warm air will open those close-set scales, and the winds will shake out the folds of little silk stamens, the catkins will be tassels of fine floss, also covered with golden powder. Nature proudly calls them flowers, and is as vain, no doubt, of these first little nurslings of the year as she is of iris, or roses, or sunflowers.

Perhaps if we have gone wandering along by some February brook we have come at last to a low, level land where the brook has so often overflowed that it has almost created a swamp. There are tussocks of grass and plenty of prostrate, nearly decayed, moss-covered trunks to walk upon. Yonder we see a large, reddish purple cone, perhaps several. Let us go close and admire these plants the more the less we touch them. "Hermits of the bog," good Thoreau called them, and wrote that they were a lesson in cheery courage to all grumblers. This thick, lurid bud is a great, fleshy leaf-like spathe. It is wrapped together much as a girl wraps a shawl or large kerchief over her head, with a point hanging down above her brow. This coarse, purple, mottled spathe looks very little like the snowy hood of an arum, but the two plants are cousins. There are hoods and hoods, the dark, torn, soiled hood of a crossing-sweeper, and the dainty, fluffy, white hood in which Miss goes to ball or opera. In Italy plants very like our "bog hermit" are called by the people capuchins, because they think the dark, bent hood looks like the head-gear of a Capuchin monk. Around the stems of these dull purple spathes we see thick, withered leaves, almost leather-like in toughness. These are last year's growth, and no doubt it is due to their protection and warmth that the sturdy, new buds push their way up at the beginning of winter, sur-vive the cold and the storms, and are ready at the first hint of February mildness to lift their heads in the sunshine. After a while, inside this unhand-some sturdy cap will rise a club, like that well-known Jack-in-the-pulpit, upon which will grow tiny blossoms; then, too, will come the true leaves of the plant, almost two feet long and brightly colored. What is this curious thing? Unluckily its terrible odor has given it its name. It is the skunk cabbage, a sharp, acrid-juiced, slightly poisonous thing, which bears are said to relish as "early greens."

CHAPTER III

THE HOPE OF YEARS TO COME

MARCH

" Up rose the wild old Winter King
  And shook his beard of snow :
' I hear the first young hare-bell ring,
  'Tis time for me to go !
Northward over the icy rocks,
  Northward over the icy sea,
My daughter comes with sunny locks,
  This land's too warm for me !' "

GOING out for a walk on some March morning, we
find the air soft and warm, the skies of a summer
blue, the water rippling in every little runnel. We
look about, half expecting to see a bluebird perched
upon a fence post, a robin stepping among the
stubble. The stems and branches which appeared
dry and dead all the winter have now a fresh exhi-
bition of life. We can almost see the sap creeping
up through their vessels and distributing vigor where
it goes.

Looking toward distant trees, their tops seem in a
single night to have thickened; they have a dim,

43

cloudy veil, instead of those sharp outlines which have stood so clearly against the winter sky. Under shelter of the fences and along the edges of the streams, even where ice needles and little heaps of snow still linger, we see green leaves of chickweed and shepherd's purse, blades of grass, fresh plants of cress.

About the lawns, upon the sod that has grown sere and brown with cold, fresh, bright tips and blades of grass come pushing up. We look at the shrubs and find the buds upon the stems enlarged, the glossy scales loosening; while the lilac, bolder than the rest, already presents an edge of green.

The world is waking up; the hosts of nature's flower-children will soon be marching over the land, while blackbird, cuckoo, oriole, thrush, robin, and bluebird sound their welcome. The coat of the bluejay is taking a new vividness; great V-shaped bands of wild-geese go honking northward; the squirrel comes from his hole; the woodpeckers whirl around the tree boles, and rap, tap, and chuckle with increasing gayety.

Where will the earth find the garments of praise, the garlands of joy, with which she will be made glorious by Maytime? They await her, packed in seeds of all shapes and sizes.

Many of us have seen a magician entertaining an

audience by taking from a hat an entire wardrobe, or from a modest little box a flowing gown, a trailing banner. Nature performs these marvels for us bona fide—not with tricky pretence. Here is a round mustard seed, from which a tall herb shall spring; from this other winged atom will come the broad, golden disk of the dandelion, where butterflies may sit to drink honey, or the tiny golden finch may rest and sway, as a fairy bird upon a fairy flower.

Jewels may be of great value, but their intrinsic worth is far less than that of seeds. All the jewels of the world might be utterly destroyed, and no more ever be found, yet the world could go on with as much health and happiness as at present. If all the seeds now upon the earth should be suddenly destroyed, and no more formed, in less than a year every animal upon the earth would be dead of starvation; within two years scarcely a living thing would be found upon our globe. These seeds at which we glance so carelessly are the hope of the world's life.

Let us go out to the woods to some sunny southern slope where maples grow. Turning over the light soft earth near the tree roots we shall find the maple seeds that ripened last autumn, and are now germinating. The seeds of the maple are in pairs, which are called keys. They look more like little tan-colored moths than keys; the distinctly-veined, winged husk is very like the narrow and veined wings of many moths.

These seeds are winged in order that they may be blown abroad on the wind and plant new forests farther afield. If they all dropped close under the shade of the parent tree few would live beyond a year or two.

Where the wing-like husks come together there is a thickening of the base of each into an ear-like lobe, holding a seed. The wrapping of this seed softens, the seed enlarges as the embryo within it grows, the husk is pushed open, and slowly comes forth the baby tree, composed of two leaves and a stem. These two leaves, though very small, are perfect, and even green, in the unopened seed. They are not shaped like true maple leaves, but are narrow, strap formed, with but one vein. They are soft and fleshy; in fact they are pantries, full of food, ready for the weak little plant to feast upon until it is strong enough to forage and digest for itself. Every one knows that babies must be carefully fed on delicate food until they get their teeth. The baby plant also needs well-prepared food.

These two leaflets are neatly laid one upon the other, and carefully rolled up, so as to occupy the least possible space. Once unfolded you could never double them up and lay them back in their case. Nature's fingers are not so clumsy as ours.

Between the two leaves is a little white stem. The two leaves unfold, and in a few days the air and sun have made them bright green. The stem between them thrusts a little root into the earth; this root is furnished with hairs. When the root is well formed and the two seed leaves have reached full size, a bud has formed in the axil between them. This is the growing point of the new tree. This bud presently opens into a pair of well-formed maple leaves.

As these leaves increase the seed-leaves diminish; the plant is feeding upon them. The ascending stem presses its first pair of leaves upward, forms between them two more, and then two more, and thus on. Small branches are formed by the end of the summer, the seed-leaves are exhausted, and the plant is doing its own work.

It is a good plan early in March to fill a box with rich, moist earth and plant in it several seeds each of peas, beans, flax, morning glory and corn. Mark the places of each kind, and take up a seed at a time during the process of growth, to mark the changes, and leave one seed of each to become a well-established plantlet. If the box has not more than three inches of earth in it, and a clean slab of white marble is laid under the earth, in the bottom, the growing roots will trace strange figures upon it, which you can see when the box is emptied and the marble is washed.

Under the trees in March we find many interesting examples of seed-growth. The feeding or seed-leaves of the young plant are called cotyledons. All flowering plants have cotyledons; the plants whose leaves have dividing or radiate veins, and whose stems are woody, or at least not hollow, have two cotyledons; grasses, reeds, corn and other grains, lilies, bamboos, all plants with hollow stems and the leaf-veins parallel have one cotyledon, while pines and trees of their class have from three to twelve cotyledons, always set in a circle.

The seed leaves of the pumpkin and squash grow in pairs, oval-shaped, not very thick, and have a sweetish taste; they have stored up sugar for the baby plant. An odd thing about these cotyledons is, that having fed the new plant until it has a good root and plumule, or top, and a little pair of tendrils, the cotyledons, instead of dying, change their shape, and become regularly formed leaves.

The acorn affords us a nice study in plant growth. Soak an acorn, peel it; you see a seam about the nut, lengthwise; split it here with care; at the tip you find a pair of tiny white points; these are the sprout or new plant. The force of life in these is

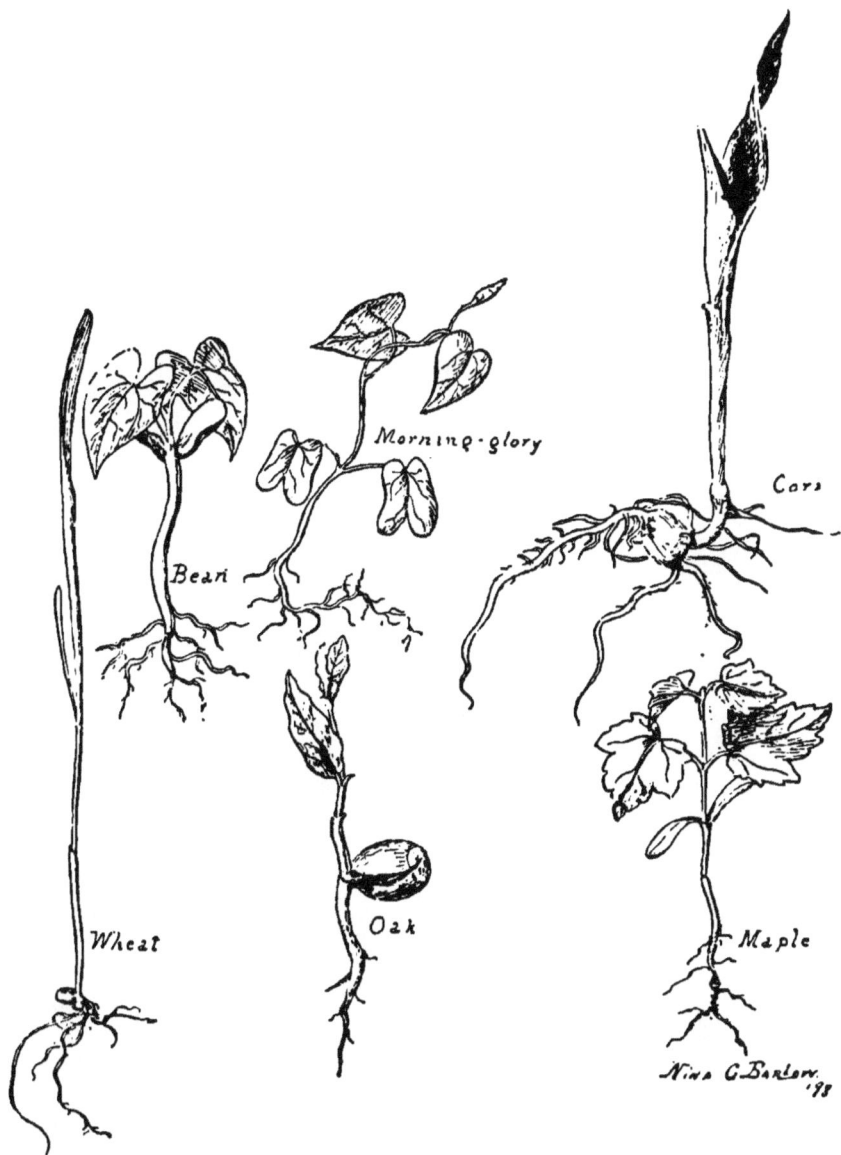

Morning-glory

Bean

Corn

Wheat

Oak

Maple

Nina C.Barlow. '98

THE INFANTS OF THE YEAR

so strong that it can split the acorn apart, husk and all; then behold! the thick white halves of the acorn meat are a pair of cotyledons! When the sprout lifts these above ground they become green; they have a large food supply for the new-born oak, and last until thrifty root and leaves render the coming tree independent.

The big horse-chestnut has even greater food-stores than the acorn. Acorns and horse-chestnuts can be sprouted in pots to study.

In peas, beans, acorns, almonds, and many other seeds the food store is in the seed-leaves, and so is a part of the young plant itself. There are other plants where the seed is full of albumen, or starch stuff, for food, and the cotyledons absorb it by their whole surface. Take, for example, a morning-glory seed. It is very hard, but as it begins to sprout this hard matter softens to a pulp, within which lies the embryo, a tiny stem and two small leaflets, or cotyledons. These lie against the store of food stuff, and eat it up before they emerge from the husk. The two cotyledons are broad and thin, with a notch at the end; this distinguishes them from the true leaves, which are heart-shaped, with a long, pointed tip.

Several times, in cutting open a lemon, I have found a seed sprouted, a stem over an inch long, and two narrow cotyledons of a bright green inside the

4

lemon. This bright green produced far from light within the unbroken skin of a lemon that had for weeks been packed in tissue paper, cork shavings, and a box, was very remarkable. In this case the juice of the lemon had supplied all needed moisture for germination.

The food of the lemon plantlet is in the thick cotyledons formed of the halves of the seed. Such cotyledons are strong, and when planted in earth are able to come to the surface uninjured. The cotyledons of the morning-glory, on the other hand, are very delicate, and having fed on the small amount of food stuff laid up in the seed, they would be broken as they emerged from the ground were it not that they remain folded flat together with the seed husk, cap-like, over their united tips. This cap serves them as nails serve the toes of digging animals, or as the hard cap at the tip of the rootlet serves the rootlet, protecting it from injury as it pushes through the earth. When these cotyledons reach the surface the sap expands them, they cast off the husk-cap, and spread apart. There are no prettier seedlings to study than a flax plant, beech-nut, and morning-glory.

The grain of Indian corn gives us another style of germination. The lower portion is soft and floury, of a sweetish taste; the upper part is almost horny. As the corn softens and begins to germinate you can

see the swollen embryo or germ lying against the
food stuff, which is the larger part of the seed. The
embryo has one thick cotyledon, and two slim leaves
twisted together inside this one cotyledon, which
wraps about them. When the husk is finally
broken the first leaf springs up and, unfolding, sends
up two others, which grow fast, while the first leaf
does not expand any more, but remains as a sheath
and support for further growth. The husk and cotyledon lie at the base of the stem until they are finally
dried up.

The Jack-in-the-pulpit springs from a hard, small
seed, and has but one cotyledon. This seed-leaf has
but a small amount of nutriment in it, and that little
is soon exhausted. Now we must remember that
Jack possesses a fleshy corm or underground stem,
which prepared food stores during the previous
summer.

Before the one seed-leaf perishes the little roots
about this corm are bravely at work to give the
plant a fair start in life, sending up its big, glossy,
green leaves. Then the leaves, the corm, and the
rootlets, all working in harmony, are able to provide
food enough for the large, thick top and its great
club-like flower cluster. The green leaves of the
arum, or Jack-in-the-pulpit, come early in the spring,
in advance of most herbage. They would be de-

voured by grazing animals, and the plant thus killed, were it not that they are full of an exceedingly unpleasant, stinging juice, which no animal but a donkey can abide.

During the first weeks of its life the arum makes leaves only, for leaves are needed to secure circulation, digestion, and breathing for the plant. At this time the corm underground is firm and plump. When the seeds are formed the corm is flaccid and shrunken, and the leaves begin to dry and fail. Their work is done when seeds are secured.

All seeds are provided with food in some form for the future plant during the early days of its growth. Starch, sugar, what is called albumen, or "white food," are laid up either in the seed-leaf part of the embryo or free within the husk of the seed itself. This is needful, because a new plant can no more obtain food for itself than a new baby, or a chick still in the egg. The human baby is fed carefully prepared food; the chick in the egg feeds upon the white of the egg that surrounds the yolk, until at last the little downy chick can peck its way out of the shell, toddle about, and forage for itself.

The seeds, the new plants or seedlings of any variety are very numerous. This is needful, as they are subject to many disasters. They may be eaten by animals or birds, decayed by overmuch moisture, withered by too great dry heat, devoured by worms, frozen, choked by too close plant growth, or ruined by overmuch shade. If plantlets were not very numerous the varieties of plants would presently die out.

This vision of the desolation of the world if the general death of the new plants happened, sets us to a busy thinking. Each year the harvest of the previous years is nearly expended by the time the fresh supplies come. The world cannot turn a Bishop Hatto and store up its corn; supply and demand, getting and using, pretty nearly balance each other. China has floods, or India has droughts, and China and India are presently starving, because last year's food is eaten up and this year's food failed to grow. When the potato crop perished in Ireland there was a famine never to be forgotten, because each summer only provided potatoes enough for a year, and when the blight cut off a crop there was nothing to fall back upon. The dependence of the greater upon the less, of the animal upon the vegetable, of man upon his plant neighbors, is impressed upon us when we are told that after that year or two of famine the Irish race never fully recovered the vivacity and easy gaiety which had until then been theirs.

When March winds shake out the leaf buds and the seeds in the ground begin to stir with strong life, we are led to think of the plant's host of enemies. There is a war on; animal fighting vegetable, and the animal would conquer in the strife, and destroying the plant would insure its own destruction, if the animal world were only at peace with itself. The army of the plant's animal enemies is an army of different nationalities, full of mutual hostilities, and its divisions are constantly turning against each other. This quarrel among its assailants brings some aid and comfort to the attacked vegetable. Some one writes in this fashion: "Nature seems to have detailed a bug for every root; worms to build nests in every tree; other worms to devour every leaf; insects to attack every flower; army worms, cut worms, all kinds of worms, and grasshoppers to eat up everything that is left. The number and variety of pests connected with each vegetable are alarming; potato-beetles come in hordes after potatoes, and fourteen distinct worms are detailed to make war on cabbage."

When in March we are digging about the roots of our flowering plants, or are having the vegetable garden "spaded up," we come upon the advance guard of the army of the plant's enemies. In the sunshine of some unusually fine March day we see an innocent-looking white butterfly sailing about— that is one individual out of the flying squadrons that

are to bring on the war; if we go deep into the ground
we shall find the " sappers and miners " at work;
there are insects to attack roots, leaves, stems, fruits.
When we look into parks, gardens, orchards, we shall
get a hint of this war. Here are trees provided with
belts of tar; other trees wearing girdles of metal,
troughs filled with water; others dazzling with a yard
or so of whitewash; yet others, where the earth at the
roots has been turned over and sprinkled with lime
and ashes. What is all this about? All this means
effort to prevent some small, helpless-looking little
worms or caterpillars from traveling up those trees
and eating all the leaves. We see the stems of young
trees cut and channeled by an insect called a borer;
others have all their once pretty little twigs swollen,
rough and black, doomed to death, because some insect has seen fit to lay her eggs in them. These
enemies of the plant will not all begin their work in
March; they are enlisting, drilling, and furnishing
their regiments. By June they will be in good
marching order, a well-disciplined host—only for
the inveterate hate some of them cherish toward
others. In March the insects are just preparing to
wake up from winter sleep; we will find plenty of
larvæ in the ground, perhaps snuggled up in the
very roots of the plant they mean to attack when
the time comes. Just now a shining, motionless

brown chrysalis looks a very harmless thing ; we cover it up when we have laid it bare, and we say " let the poor thing sleep ;" the poor thing will wake up from its nap and try to rob the big world of its dinner.

Our house-plants, kept warm and moist, nourish their enemies into an early growth. In March we see some of our favorites growing yellow and hang-ing their heads. If we have good eyes or a strong lens we shall soon see what is the trouble ; here are detachments of " red spider," not so big as the head of the tiniest pin, but with a great appetite for leaves, which he riddles and nips and kills, in spite of his tiny dimensions. On another plant an amaz-ing colony of aphis has settled down ; a very wonder-ful little bug, the aphis, about which learned articles, and even books, have been written. It looks a help-less mite, not much larger than a poppy seed, and dressed in green like the leaves it lives on. Unless we can get rid of the aphis we shall soon lose our plants. Our best ally here will be another of the insect army, a beautiful, friendly little creature that never harms man and his plant partners and pro-viders. This is the lady beetle, or lady bug, a dainty creature, dressed in red, black dotted, or in orange or black with red dots.

Out of doors that delicate white butterfly, drifting

up and down on the breeze, fair as the soul of a flower, will lay an egg out of which will crawl the cabbage-worm, terror of gardeners. Over the gooseberry and currant bushes a host of simple, transparent-winged flies will swing up and down in early May. Just now these flies are wrapped up in snug cases in the ground, so well hidden that we cannot find one of them. In May, as they swing about in the sunny air, they will seem to grow weary and will keep darting upon the bushes; fatal enough are those seeming restings. The saw-fly is not as careless as she appears to be; she is laying her eggs, and swarms of currant-worms will come forth to strip the bushes of every leaf and destroy all hopes of gooseberries and currants, unless the legions of the worms can be killed by poisonous sprayings.

The chinch bug, the squash bug, the Colorado beetle, the grasshopper, and the tent-caterpillars are now, in March, having the final "beauty sleep" of their winter rest. If they and their fellow-soldiers can have it all their own way in a few weeks there will scarcely be a plant left to comfort our hearts. But right along with these insect enemies, among the roots and under the bark of the plants, hidden in cozy nests in axils of buds and stems, swinging in woven cradles on twigs, or the under side of dry leaves, are the deserters from the ranks of insect enemies, the revolting squadrons that will fire upon the battalions of their own armies. Here are the larvæ and pupæ of the black ground-beetles, harmless, helpful things, that we think so ugly because their legs are so long. Here are the cradles of the splendid tiger-beetles, enemies of all spiders. Wheelbugs and soldier-bugs, asleep under grass in tiny hard cases, will come out in a hurry to save the wheat crops. All the beautiful lace-wing flies that spend their infancy among the cool shadows of water-plants, and are now watching the stirring of spring life among flags and rushes, will come to the rescue of the plant world, because they have big appetites, and nothing satisfies them but the juices of other insects.

Here in March, as our crocus blooms push above the brown mold, and the hyacinths and tulips send up points of green, while the peony reaches up crimpled leaves red as rubies in the sunshine, the greatest of all the allies and the defenders of the plants have come to welcome their earliest awakening—here are the birds. Why were not all the plants destroyed when there were no farmers and gardeners around to sprinkle and spray and pick and shake and net? How comes it that the wild plants and fruits are not driven from the face of the earth by insect enemies? Why? Simply because in these wild conditions of nature's making the birds are not killed, frightened or driven off, and for every berry or bud a bird may take for his food, he saves perhaps a million. It has been estimated that the birds of Nebraska in six months eat up five billions of insects, and would clean up of noxious insects over eighteen thousand acres a day.

When in March we dress up our garden borders, or walk out to watch plant growth in the woods, we see nothing of the harmless silent toads, among the kindliest friends of the plant. The little gray toad and the small, prettily colored ground snakes live entirely on animals that are harmful to plants—as worms, insect larvæ, mice, and moles; the little snakes taking these last in hand—or rather in jaws. The one fault of the little garden snake is that he likes to eat toads. The toad, on the contrary has no faults at all. When he wakes from his winter's nap he will sit down by some plant and pick off, one by one, every insect enemy that comes intent on evil.

CHAPTER IV

APRIL

"Summer is y'coming in,
    Loud sing cuckoo !
Groweth seed, bloweth mead,
And springeth the weed anew.
    Sing cuckoo, cuckoo !
Well singest thou, cuckoo,
Never thy like I knew;
    Merry sing cuckoo !"

FOLIAGE is the most prominent feature of the plant world. Trunks and branches are large and grand; the parti-colored flowers are, at first glance, more beautiful, but the leaf is the most conspicuous part of vegetation. If flowers and leaves exchanged places, and wherever is now a leaf we should have a blossom, the eyes would soon tire of the glare of vivid color, and we should long for the soft, restful green of leaves.

When we speak of a leaf we think at once of the flat, green, expanded body, springing from the stem, or from the hidden root crown. This is truly the

60

typical, or chief form of the leaf, and when there is a departure from it we call such departure a "modified leaf."

The white thick scales of the onion or lily bulb are leaves ; the scales which wrap up buds, the cotyledons, the thick divisions of the house-leek, the tendrils of the pea-vine, the thorns of the barberry, the brown woody scales of the pine, or larch-cone, the needles of the pine, the little, odd-shaped bracts on stems, are all leaves. In fact, all the parts of the flower, the calyx, petals, stamens, and pistils, are really leaves. When we now speak of leaves, we mean leaves in the popular sense, the common leaf, which is the type or pattern of leaves usually borne on a stem, a green, expanded body, performing for the whole plant various and necessary offices.

Early in April we find the leaf buds unfolding upon the sides of the stems, or pushing up through the ground. Some of these buds are placed opposite to each other upon the stem, some are in rings around the stem, others are set alternately, others spirally, so that if you follow with a thread the placing of a certain number of buds you will see that the thread has made a complete circuit of the stem, and then another. Where the leaves are in a spiral placement it is merely a whorl drawn out; where there is a whorl it is but a compressed spiral.

On pines the needle-formed leaves grow in bundles
or clusters, and a scale-covering wraps all the buds
of such a cluster together; bundles of this fashion
are flattened branches.

Having noted the placing of the buds, let us see
how the leaves are packed away in them before
opening. The leaf of a cherry or oak is folded flat
together by the mid-rib, or woody vein in the
centre. A currant leaf is folded like a tiny fan furled
up; a plum leaf is rolled up inward, first one side
being rolled, and the other rolled over it, toward the
mid-rib. Some leaves—as those of the azalea—are
rolled over backward to the mid-rib on each side,
making two little tubes. Exactly the opposite is the
manner of rolling of the violet leaves, where the
tubes lie on the upper side.

When the leaf buds are nearly ready to open, you
can study the method of packing if you have a microscope. Cut a bud across the thickest part, and examine the cut sections.

Let us consider the great number of leaves that
each plant possesses—the countless blades of grass,
the multitudes of pine-needles, the heavy shade of
forest trees, produced by small leaf overlapping and
overhanging leaf in a vast, almost impenetrable
dome of verdure. We wonder if there can be more
drops in the ocean, or more grains of sand on the

seashore, than leaves on plants; yet every leaf has had all this careful folding and placing in the summers that are past, and untold hosts of leaves are now being placed and packed for summers yet to come!

Leaf buds are generally protected from wet and cold by thick scales, fine down, hairs or coats of a resinous gum, exuded from the plant. These protective coverings give way when, by the rising of sap and the effect of warmth, the buds begin to expand from within. The first days of April show us the swelling buds; by the end of the month the world is in a gala dress of expanded foliage.

An ordinary leaf consists of a blade, or broad part, a footstalk to hold it to the main stem, and a pair of stipules or wing-like, green expansions, on each side the footstalk. Stipules are well shown on rose and clover leaves. Some leaves have no footstalk, many leaves have no stipules, therefore we see that the blade is the only really needed part of the leaf.

Let us look at a leaf blade. The woody fibre which makes up the main stem and, bound into a little bundle, composes the footstalk, spreads out into a light, woody framework for the leaf. This framework is usually in two layers, like the nervures in a butterfly's wing. The central line of the frame is called the mid-rib, the other parts are styled the veins. Some of these veins are coarser and stronger than others, as for example, those which expand in the large side-lobes of maple and oak leaves ; other veins are as fine as spider's web. Every student of botany should make studies in venation, by soaking leaves until the green part has decayed, then laying them on black cloth, and brushing the pulp away gently with a fine brush, when perfect specimens of frame-work will remain.

The form of the leaf depends upon the woody frame-work. Upon it is spread the green pulp called parenchyma. This is composed of cells, while the woody frame-work is formed of expanded or modified cells, called vessels. Over all the leaf is laid a thin, transparent skin, named an epidermis, which covers the footstalk, stipules, and all leaf appendages. Usually the epidermis on the upper side is more glossy and of a finer texture than on the lower surface, which is softer and more porous. As we shall see, there is good reason for this.

All plants which have one cotyledon have hollow reed-like stems, and their leaves have parallel veins ; that is, veins running side by side, without branchings ; grasses, grains, rushes, lilies, palms, bamboos, are of this one-cotyledoned, parallel-veined family.

There are two different kinds of parallel venation. There may be a mid-rib, and the veins run from this to the margin, as in the calla lily and the pickerel reed. The other fashion shows no mid-rib, and the veins all run from stem to tip, as in the corn, grasses, lily-of-the valley, and others.

Some net-veined leaves appear at first glance to be parallel-veined; a little examination will show the difference.

Plants which have two seed-leaves, or cotyledons, have woody or not hollow stems, and the leaves have radiated or net veins. Reaching the beginning of the blade, the woody fibre in the stem divides into a main central line and various branches. These veins are not only a frame-work, but as they are hollow they serve as canals or a circulating system to carry the sap into all parts of the leaf and return it to the stem, to visit all other parts of the plant.

The form of the leaf in even minute particulars depends upon the carrying out of woody fibre. Whenever it ceases or gives way the parenchyma or pulp and the covering skin also give way. In this manner the great variety of form in leafage is produced.

All parallel-veined leaves have smooth, even edges. Some net-veined leaves, as those of the senna, the buckberry and American indigo, have also even edges. Most net-veined leaves have cut or uneven margins.

5

In some leaves the edge is waved in large scallops, as in the nasturtium; others have smaller, rounded scallops: some are edged in sharp points, setting straight out or forward or backward. Some leaves are more or less deeply lobed, like those of the maple, oak, currant; some look as if pieces had been deeply cut away, almost to the mid-rib; others are slit or cleft, but not cut out. The castor-bean has its leaves divided entirely down to the mid-rib. Some leaves have the lobes spread out like an open hand; others are divided like feathers.

A compound leaf is a large leaf made up of many little leaflets, as those of the locust and rose. The horse-chestnut leaf has five divisions, the clover three; the meadow rue, maiden-hair fern, and honey-locust have many. Sometimes, as in the pea and vetch, the end leaflet of a compound leaf is turned into a tendril or two for climbing purposes.

· The footstalk holding the leaf to the main stem is generally short; sometimes it is very long, and used as a tendril for a climbing plant, being wrapped around and around some support. Some leaves are sessile or without footstalk, sitting close upon the main stem, or the branch. Other leaves seem to have the stem thrust right through their lower portion, as in the wild bell-wort. In such case the lobes of the leaf have grown together, clasping the stem. The coral honeysuckle appears to have the stem growing through the very middle of each leaf; in this plant two opposite leaves grow together at their lower portions.

Some leaves are heart-shaped, as those of the morning-glory and violet; others are quite round, as those of some water-lilies; some are lance-shaped, or arrow-shaped, and so on for many varieties. The student of botany should draw or press examples of all these shapes and margins.

While all these forms of leaves have some resemblance to each other and their position as foliage is clearly evident, other leaves are so remarkably modified that their character is scarcely recognized at first sight. Thus the barberry rolls up some leaves into sharp, woody thorns, the mid-rib embracing all the other parts. The pea, as we have seen, details some leaflets for tendrils; the agave, or century plant, builds up great leaves, one or two inches thick, using the green outer portions to do leaf duty, and the white inner part to serve as storehouses.

The sun-dew and dionea, or fly-trap, separate a portion of the leaf, edge it with prickles, provide it with honey-drops, turn the mid-rib into a sensitive hinge, and lo, a bait and a trap for catching unwary insects! The pitcher-plant in its several varieties unites the outer edges of a part of the leaf into a water-tight pitcher, which it keeps full of liquid.

The general color of leaves is green, but some, as those of the coleas, are of various brilliant colors — scarlet, yellow, red, purple; other leaves, as some varieties of bergamot, begonia, striped grass, have green leaves, mottled with white, yellow, or brown; other plants have green leaves and parti-colored edges.

Some leaves emulate the blossoms in fragrance; the orange, lemon, mint, sage, thyme, geranium families have in their leaves a rich odor. All this variety in form, color, fragrance, adds greatly to the beauty of foliage. Ruskin says: "The leaves of the herbage at our feet take all kinds of strange shapes, as if to invite us to examine them. Star-shaped, heart-shaped; fringed, fretted, cleft, furrowed; in tufts, spires, wreaths; never the same from foot-stalk to blossom, they seem perpetually to tempt our watchfulness and to take delight in outstripping our wonder."

Leaves were not created chiefly for beauty, but for use. What is the use of the leaf? "To give shade," you say. That is one, perhaps the least important of the functions of the leaf. Animals and plants alike are indebted to the shade of foliage for much comfort, and for some further possibilities of life and growth.

You suggest, as another use, the supply of food. Yes, the grasses and many herbage plants are greedily browsed by animals; we owe to them our supply of beef, mutton, milk, butter, cheese, and other articles of food. Lettuce, cabbage, cress, parsley, are familiar examples of leaves eaten by man, while endive, chicory, thyme, sage, dandelion, mustard, are more or less used upon the table.

We have not yet reached the most important functions of the leaf. To the plant itself the leaf serves as a food purveyor, gathering perhaps the larger portion of plant food from air and moisture by absorption. The leaf is also the main breathing apparatus of the plant; the leaf spreads out to air and sunlight the food matter received by the entire plant, and thus secures chemical changes in it similar to assimilation and digestion. The leaf makes possible the circulation of the sap. Thus the leaf serves the plant as throat, lungs, and stomach. What the human being would be without such organs the plant would be without the leaf, or some part modified, as in the cactus family, to serve the purposes of the leaf.

How does the leaf perform its duties to the plant? The root absorbs water, holding in solution various mineral substances; this rises as sap through the tubes of the stem to the leaves. There it is spread out through the veins to every part of the leaf surface. Thus it is exposed to light, air, sun-heat, until it is chemically changed, cooked, as we may say, into good plant food. The cells in the parenchyma hold the coloring matter of the leaf, which is called chlorophyll, or green stuff. This is the chief agent in digesting plant food. There is also other coloring matter known under one general name of chromule. The presence of sap in the leaf adds especial lustre to these coloring materials. In spring, when sap is abundant, the leaves are much richer and brighter in color.

When leaves have fallen in the cold season the roots are also at rest in the soil, the stem ceases to send up sap; the whole plant seems asleep. You may cut a sugar-maple stem in winter and no sap flows; if you cut it in spring the sap rushes out. Grapevines and trees should be pruned in the late autumn, so that the scar may harden over before sap rises in the spring. This care will prevent leakage, weakening to the plant.

In the potato and other plants having tubers or corms, a large portion of the ascended sap, laden with the food-stuff gathered from the air by the leaves, is sent back through the stem vessels to the underground storehouses. When, as in the agave and other like plants, the leaves are themselves the storehouses, a large amount of food is retained in the fleshy tissues of the leaf.

From the air the plant absorbs an abundance of carbon, with less oxygen and hydrogen. From these, and the material brought up by the sap from the soil, the plant, chiefly by the agency of the leaves, makes starch, sugar, resin, gum, oils, jelly, and a great number of substances useful to men, and eventually becoming the food-supply of the animal kingdom.

Leaves receive constantly more or less moisture from the air. This is chiefly absorbed by the under surface of the leaf. Leaves that hang sidewise, having the edges to earth and sky, are alike on both sides; leaves that are set horizontally, with the one surface to the earth, have that surface softer and more porous.

The leaf surface being full of cells, exposed to air and heat, their sap contents become rarified or given off, and this causes an upward pressure of sap through the stem-vessels to fill them; these, being crowded full, the sap, by natural gravity, begins to seek the lower parts of the plant, and thus two constant currents of circulation are kept up—a bringing up of material, a carrying of this material changed to food-stuff downward, and a distribution of it for building purposes throughout the whole plant.

The foliage not only prepares food fit for plant use, but it also prepares air fit for the breathing of animals. The out-breathing of all animals is loaded with carbonic acid gas, which is a poison; their in-breathing takes from the air oxygen, which is wholesome and valuable for all animals. It is evident that all the air in the world would become loaded with carbonic acid gas poison and robbed of useful oxygen were not some way contrived to exhaust the one and replace the other. Here the plant comes in to restore the balance of affairs. The out-breathed carbonic acid gas is the chief food of the plant world. Oxygen is a drug in the market to plants; a very small trace of it suffices for their needs. Thus while men breathe out carbonic acid gas and breathe in oxygen, plants do exactly the reverse; they out-breathe oxygen and in-breathe carbonic acid gas. Thus the plant and the animal form a mutual benefit society, and prepare air that suits all concerned. The green parts of the plant, chiefly the leaves, are the agents in this constant reconstruction of the atmosphere.

Activity for many months in all these directions exhausts the energies of the plant for the time being. The leaves, the stem, the veins become clogged by the abandoned particles of solid matter. This is especially the case in the leaf, stem, and veins. The seeds being perfected, the work of the plant for the year is accomplished, and the root-mouths are less active. The capillary vessels and the leaf-cells no longer force sap through the plant. Unfed by sap, too much clogged to feed from the air, the leaf fades. At the axils of the leaves the buds for next spring are formed, and thus by surfeit and by pressure the old leaf is severed from the parent stem. This is rendered more easy by a peculiar construction of the cells at the foot of the leaf-stem.

Properly to perform their offices, leaves need abundant light and air. If they grow under water or in dense hedges, where the supply of air and sunshine is limited, new plans must be laid for securing what is indispensable. We find that leaves growing in such situations are very finely divided, even when in other conditions the plants bear leaves but little divided. Where great economy is demanded, at once the plant forms deeply-divided leaves which shall act as a seine to intercept as much as possible of the light, air, and moisture demanded for converting crude sap to plant-food. The additional surface secured by these repeated divisions makes up for the harder circumstances of their lives.

Organs so valuable as leaves must be protected. Leaves are sometimes defended by sharp prickles, as in the cactus, thistle, holly; by rough coats, as the

mullein; by tough epidermis, as in the plantain and
yucca; by pungent juice, as Jack-in-the-pulpit; by
nauseous taste, as the burdock; by height, as the
palm, oak, cedar. Some hide under water, some
float on ponds, while mere numbers secure others,
as the grasses.

However much interest we find in the varieties,
the development and offices of leaves, it is impos-
sible to confine our attention strictly to leaves when
in April we carry on plant-life studies out of doors.
As March closes, flowers are scarcely to be found,
but the first breath of April seems to have awakened
a host of the darlings of the spring. Perhaps the
earliest hint of a flower is when one finds the earth
strewn with small, red scales and tiny clusters of
filaments. We look up. The tops of the elms against
the blue sky have suddenly thickened, these scales
and threads are the bloom of the elm tree. Looking
from what is high to what is very low we find the
ground almost covered in places with the tiniest of
flowers, the dainty little bluets; on stems half an
inch long, surrounded by leaves no larger than a
small grain of rice, these wee blooms, not much
larger than a flax-seed, but perfect in form and
matchlessly blue in coloring, too tiny to pick or to
put into bouquets, challenge our liking, small babes
of the opening season.

Under the fences or on the wet sides of runnels along the road the chick-weed opens its white stars. Many of the spring flowers are white. Conspicuous among these are the blossoms of the shad-bush or service-berry, sometimes called "June-berry," because its clusters of red fruit ripen in June. The shad-berry is a tall shrub growing on low hillsides near brooks, which it seems to delight to overhang, viewing its graceful reflection in the still mirrors of water between the stones. The snowy-clustered blossoms come while the tree is still bare of leaves, and every stray breeze tosses about the long, slender loosely-hanging petals, which seem at first sight just ready to be blown away. This plant has one of its names from the notion that its white banners were unfurled when the first run of shad began to ascend the rivers. Early settlers named it also "service-berry," from the use which the Indians made of the fruit—drying it, beating it into a kind of cake, or squeezing out the juice for a drink.

The white blossoms dance on the April air, timed to that refrain so glad to flower-lovers, "Lo, the winter is past." Do the beautiful wood anemones hear and understand that? They are the next flower that appears in the spring. Now and then one or two come out in the last days of March, but they are small and pale, seeming to shake with the cold. All the poets have loved the anemones. Bryant tells how "gay circles of them danced on their stalks," and Whittier sings of how "dainty wind flowers sway." White is the usual dress of the anemones, but we have seen them pale purple, pink, or blue. One oddity about them is the variable number of their petals; one may count them all the way from four to fifteen : probably five is the normal, and when more appear some of the stamens have perhaps chosen to abandon their pollen-bearing and expand to the more showy petal form. These petals also are sometimes dissimilar in shape, in a five-parted blossom each segment is nearly oval; where the petals are numerous several are queer, crooked, crimpled, triangular, or cone-shaped affairs. The foliage of the anemone is as delicate and pretty as the flowers, usually three leaves, each made of three deeply-serrated leaflets, from among which the simple stem lifts the frail, solitary blossom. This wood anemone has a cousin, the rue anemone, so called because its leaves resemble those of the rue-plant. This rue anemone has smaller blossoms than the wood anemone, several springing on short stems from the same axil as the leaf stalks. The rue anemone likes a deeper shade than its cousin, and loves to hide at the roots of old trees.

We have spoken of the " corolla " and " petals " of

the anemone; •this was, in fact, a concession to appearances, the white petal-like segments are really sepals; there is no corolla, the calyx is the whorl of showy-white or colored sepals surrounding the stamens and pistils. The root of an anemone would have afforded a pleasing specimen for our study of roots—it is composed, in the wood anemone, of a singular, fleshy, scimiter-shaped stock; from the point of the scimiter, or near it, rises the smooth, simple stem, from the other extremity of this rootstock branch several long, slender rootlets. The rue anemone has five tubers shaped something like a sweet potato, from each of which depends a cluster of branched rootlets. The Greeks had many stories and romances about the anemone. They believed it sprung from the tears Venus wept over dead Adonis; they gave its name as signifying that the blossom opened at the wind's bidding, and they called it "The wind shaken." The Persians said that the anemone had a subtle poison which it poured upon the wind, and the flower with them was an emblem of sickness. Perhaps these Oriental anemones were unlike the flower to which we give the name; our anemones, loving the winds and the woods, seem emblems of vigorous health, under a delicate appearance.

Another white flower, among the first to bloom in the April woods, while yet there are almost no leaves to cast a shade under the trees, is the beautiful blood-root, the sanguinaria, so called from the crimson juice of its medicinal root. The sanguinaria is a showy flower as large as a silver dollar. This snowy denizen of the woodlands is a close cousin to the gaudy poppy. The leaf of the blood-root is large, firm, deeply-lobed, and with thick veins. The leaf stem and the flower stem leave the fleshy rootstock together, emerging from a scoop-shaped scale and curving upward, the flower-stem lying within the clasp of the leaf stem, and the flower bud, cased in the two sepals of its calyx, resting within the leaf, which is closely curled about its precious charge; its firm, well-folded tip making its way through the mold. The leaf, having safely emerged to the light, begins to expand, and the flower stalk greatly accelerates its growth, lifting itself from the clasp of the leaf and becoming twice as high as the leaf stem. The two sepals bend back against the stem, and presently drop away. The silvery-white salver, composed of from seven to twelve strap-shaped petals, expands broad and beautiful, with a cluster of some twenty-four short golden stamens, crowded around a single pistil. The beauty of this flower is very fleeting, two or three days will hasten the maturity of all these spotless blossoms, and then the first breeze will sweep them all away, leaving the pistil to expand its sharp-pointed seed vessel.

The Indians used the indelible juice of the sanguinaria root-stock as paint for their faces and weapons.

These April woods seem full of white flowers, and none is more delicately lovely than the wild dicentra, otherwise known by the ugly name of "Dutchman's breeches." What a pity that these absurd or vulgar names fasten upon some of our daintiest flowers! When the warm south hillsides suddenly are covered with the waving plumes of this charming plant, we find it hard to tell which is loveliest—the tassel of white, yellow-tipped bloom, or the fine frilled bluish-green foliage. These leaves are thrice compound, very deeply divided, the edges seeming ruffled, so full is the border of the leaf. The whole plant is smooth and shining, a certain fineness and frailty characterizing every part. Each blossom has two tiny scale-like sepals; the corolla is made up of four closed flat petals with yellow tips. The outer petals are much larger than the inner two, and swell and spread at the tip, ending in deep spurs. The two inner petals are shaped like tiny spoons, and close over the stamens and pistil. A pretty name given in some localities to this flower is "White Hearts," from its heart-shaped corolla.

No wild flower is more universally beloved—perhaps the violet might be excepted—than the spring beauty. Strong, abundant, lasting, taking kindly to being picked and put into a bouquet; readily domesticating itself on our lawns, if it is only treated with so much courtesy as allowing its seed pods to ripen before the lawn mower goes over them — such a flower is our spring beauty. Some botanists refer to this plant as a lover of damp woods or brooksides, but it seems to love any spot where it may be permitted to grow. The leaves are two long, narrow straps, springing opposite each other from the base of the stem. A number of blossoms grow in the loose cluster, the top one opening latest. This flower expands in broad sunshine and closes itself under a cloudy sky. The calyx has two sepals, the corolla five white petals striped with pink, which give the whole flower a rosy hue. No one of our wild flowers is a more profuse bloomer, none lasts longer, almost none arrives earlier to tell us "The flowers appear on the earth, the time of the singing birds has come."

CHAPTER V

THE BEAUTY OF THE FLOWER

MAY

"And all the meadows wide unrolled,
  Were green and silver, green and gold,
  Where buttercups and daisies spun
  Their shining tissues in the sun."

NEITHER age, learning, nor fortune ~~are~~ needed to
enable one to love and admire these gracious children of beauty—the flowers. When the chill winds
of autumn sound a knell for their departure we have
a sense of loneliness and loss. As the winter passes
we long for the days when the blossoms shall come
again.

The first tiny white blossom of the star-flower; the
first little tasseled bloom on the birch; the first
adder's tongue, or violet, or broad white salver of the
mandrake flower; the long, white, plume-like cluster
of the foam flower; the snowy banners of the dogwood; the gray-white of the brave little plantain-
leaved everlasting, fill all hearts with delight.

The aged, little children, invalids, strong men, the

laborer in the field, the lady in her parlor, all are happier and better for the coming of the flowers.

> "For lo, the winter is past!
> The flowers appear on the earth,
> The time of the singing birds is come."

We have noted great variety of form and color in leaves; in the flower we find in these respects infinite diversity. How many children have spent hours in vainly searching for two exactly similar blades in a clump of striped grass? How many more have spent other hours in seeking for two pansies exactly like?

The life object of the flower is the production of seed. All the parts of the flower are in some way fitted to further that end. What is the story of the flower?

The stem and branches having developed a certain amount of leafage, may at length put forth blossoms. These spring, as leaves do, from the tips or axils of the branches. In truth, a flower is a modified branch, and all its parts are modified leaves. We will pass over this distinction of science, and will consider the flower as we popularly think and speak of it, the beautiful producer of seeds.

On most trees the flowers precede the leaves; there is enough sap in the stems for their production, and

NATURE'S DARLINGS

if they delayed until the leaves spread out they
would be deprived of sufficient air and light. In the
maple, elm, oak, birch, willow, and others, the
flowers come first. In the catalpa, tulip tree, and
magnolia, which are large-flowered, the leaves come
first, but the large blossoms appear on the tips of
the stems, beyond the leaves, and are not interfered
with.

On many of our ornamental shrubs, as the red-
bud, dogwood, Irish quince, and others, the abundant bloom appears before the leaves are conspicuous.

The folding of the flower in the bud is as wonderful
and interesting as the folding of the leaf, and should
be studied in the same manner.

The placing of flowers upon the stem is also various
and worthy of close observation. When flowers
grow in clusters the leaves near them are small,
modified in form, and known as bracts. These bracts
frequently appear at the base of the leaf stems of
flowers, whether placed singly or in pairs.

Where a number of flowers grow near together
they are more or less crowded, and are said to grow
in heads, spikes, clusters, tassels, umbels, catkins,
bunches, spadix, and so on.

The cottonwood, clover, and button-bush give
examples of flowers in heads; also the snow-ball bush
and globe flower are well-known specimens; the

birch and willow show the catkin style. Jack-in-the-pulpit exhibits the spadix method. The big-hooded pulpit is not the flower—it is a sheathing leaf; Mr. Jack himself is the flower. If you open the pulpit you will see that the lower part of the club, called Jack, is closely set round with little blossoms. This club-formed inflorescence is a spadix.

If you took a flower head and could pull out the axis or cushion upon which the flowers are crowded, and make a roll of it, you would have a spike growth. The weed called plantain has its flowers in a spike.

The lily-of-the-valley grows in a loose branch or raceme; the long stem springs from the folded base of the leaf, and upon it the lilies hang on tiny stemlets. The lower ones open first. If you could flatten down this raceme to a disc, you would have flowers placed on a flat, circular cushion, the outer ones blooming while the central ones are buds. You will notice in many flowers that the blossoms do not all open at the same time; the mullein and the yellow evening-primrose open the lower blossoms first; they are dead and seeds have formed, before the upper buds open.

What we call a dandelion, and a thistle flower, are really made up of many little florets growing upon a single fleshy cushion, surrounded with crowded green bracts, which form an involucre, or wrapper. Flowers growing in this manner are embraced in the great order compositæ. In such compound flowers each floret may be perfect in itself, having its calyx, corolla, stamens, and pistils. Sometimes the outer row of florets has neither stamens nor pistils, but has several petals united into a broad strap; such straps make a bright border for the whole compound flower. The florets may have stamens only, or pistils only. Asters, coreopsis, zinnias, cockscomb, golden-rod, chrysanthemums, belong to the compositæ, with a great many other flowers which each student can search out for himself.

One very large family of plants is called umbel-lata, because its flower stems all start from one point, spreading out something like the wires of an um-brella, the flowers opening at a level, instead of in a head or spike. The caraway, parsley, wild carrot, and parsnip show this very common form of growth.

A flower is simple when it has its own stemlet, whether or not these stemlets are crowded into heads or spikes. A flower is compound when many flowers are wrapped in one involucre, and appear like a single blossom when they are really many, as the sunflower and dandelion.

The only really needed part of the leaf is the blade, so the only really indispensable parts of the flower are the stamens and pistils. These, held together by a little scale, can produce seed, and some plants are so economical that they reduce their flowers to this footing.

Many trees and nearly all the grains and grasses have this inconspicuous fashion of blossom. They seem content to be taken on their merits, rather than upon appearance. If such flowers have any color it is in the stamens and pistils. Grass in blossom is very beautiful; the stamens and pistils are red, yellow, pink, purple, hanging loosely on the stem like a gauze decoration, with an undertone or ground of pale green bracts.

Only for a day or two can this simple and exquisite bloom be seen, for few things are so evanescent. "As brief as bloom upon grass," is proverbial—

"For the sun is no sooner risen with a burning heat,
But it withereth the grass,
And the flowers thereof falleth,
And the grace of the fashion of it perisheth."

"Grace of the fashion of it"—that describes the grass bloom exactly.

What is called a perfect flower we will examine in the common buttercup of the fields. At the top of the stem we find a cup or calyx of five narrow separate green leaves, called sepals; these form the outer wrapping of the bud, and maintain and protect the more delicate inner parts of the flower. Within the calyx is the corolla—five glossy, yellow, roundish petals, set in a circle; within this we have another ring of downy, bright-yellow stamens, and still within these, protected by all the others, certain yellow pistils, fewer and firmer in texture than the stamens. All these four rings of parts are placed upon the fleshy, enlarged top of the stem, which is called the receptacle. The yellow of this flower is very yellow, the stem and leaves are very green. The stem and leaves of our buttercups are hairy; the whole plant is provided with a sharp, stinging juice.

There is a great variety in the setting of these parts of the flower. In some flowers the parts all adhere to the calyx; in some the ovary or lower part of the pistil is below, or inferior to the rest of the flower, as in the rose, where we find the rounded button of seed box below the spreading green sepals. In the lily this ovary is free from the other parts, and is superior to them, standing within and above the rest.

The color of the calyx is usually green, and its first office is to maintain and protect more valuable organs. When the corolla is absent the calyx frequently changes its color, and adds to its functions, becoming bright-hued and serving to attract insects to the flower. The purpose of this bidding for insect visitors will be explained in Chapter VII.

The corolla is composed of two parts, banner and claw. The banner is the wide part; the claw holds fast to the receptacle. The stamen consists of two parts, the filament and anther. The filament is the stem part of the stamen; it may be long, short, or wanting altogether. The anther is a little box, usually oblong, composed of two lobes and opening in the centre. Within the anther grows a fine yellow dust called pollen. Sometimes the pollen is not yellow but dark-brown or reddish. When the pollen is ripe the anther opens and the pollen flies out. Pollen grains can only be examined with a microscope. We shall find that each grain is a little sac, full of liquid, in which is held a constantly moving atom; this atom is the life point of the flower.

The pistil has three parts. At the base is a little box called the ovary; then comes the style, a hollow tube bearing on its top the stigma, a soft, sticky cushion. When there is no style the stigma is fixed directly upon the ovary. The ovary is full of little ovules or egglets that will presently become seeds. These ovules cannot develop into seeds unless the pollen reaches them.

The grains of pollen must first reach the sticky surface of the stigma; being held fast there, they sprout, as seeds in soil, and send long threads down through the hollow style into the ovary. These threads are hollow also, and through them the quivering life atoms reach the ovules, which at once receive power to grow into true seeds. The whole plant then sets itself to protect and nourish the seeds. The germ box or capsule hardens its fabric, the sap brings the choicest food, and the growing seed develops all those parts which we noticed in the sprouting of the seedling—case, cotyledons, albumen-food, tiny plant embryo.

The buttercup, as we have seen, is made up of four circles, each composed of several distinct parts. A flower with several petals is called polypetalous. Other flowers have but one petal; they are styled monopetalous.* In fact, in such one-petaled flowers a number of petals have simply grown together. Let us take a morning-glory as an example. Pull off the calyx; it comes off as a whole, but is cleft half way down into five lobes, showing that it is truly composed of five united sepals. Now pull the corolla from another calyx cup; it comes as a whole, and is not cleft as the calyx is, but it has five stripes, and at each stripe the margin has a little point, and

---

* Gamopetalous and gamosepalous are newer and better terms.

we can make out very plainly that here are five prettily-pointed petals united into one, with a long tube made of the claws, and a beautiful wide margin made of the banners. Four-o'clock, stramonium, Canterbury-bells, phlox, and many other flowers have these one-petaled corollas. Such corollas differ greatly in shape, owing to the length and diameter of the tube and margin.

The snap-dragon shows a pretty variety of a one-petaled corolla, where the margin is drawn together into a mouth. The fox-glove, with its inflated throat and narrowed margin, is another beauty; and the lady's-slipper, with its yellow or purple bags, hung under the long, wavy calyx sepals, is one of the most beautiful of flowers.

In polypetalous corollas we have the rich splendors of roses, from single to the fullest double, where cultivation has changed all the stamens and pistils into petals. The polypetalous tribe give us also the lovely perfume-filled chalices of the lilies; the peas, with their many-colored banners; the charming violets, with their spurred petals; the columbine, with its horns of plenty.

Color of some kind is one of the distinguishing features of blossoms.

Fragrance is another marked characteristic of plants, and is chiefly in the flower. There are plenty of scentless plants, yet the majority are full of perfume. Some few, as the elegant crown imperial, have a very disagreeable smell. Fragrance in plants comes from certain oils or resins laid up in different parts of the plant, whether in the leaves, bark, wood, fruit, seeds, or blossoms.

Some flowers have especial glands or developments called nectaries, in which honey is laid up for insect guests. All children have sucked the honey from the lower parts of the long tubes of the honey-suckle, or clover, or from the spur of the violet.

Fragrance and nectar are especially abundant about the time of the ripening of the pollen. After the anthers have shed all their pollen the flower begins to fade, and then the fragrance loses its delicate sweetness, becomes heavy and sickly, or positively unpleasant. At this time also the nectar dries away, unless it has all been previously drunk up by insects.

A perfect or symmetrical flower is one having all its parts evenly placed and in equal number in each set of organs. For example, the flax flower is a perfect, symmetrical flower, with five sepals, five petals, five stamens, five pistils. The trillium or bath flower is divided into threes: three sepals, three petals, even three sides to the capsule, and three lobes on each leaf.

From snow to snow, from crocus to chrysanthemum, we have about us our flower guests.

"Consider the lilies, how they grow ; they toil not, neither do they spin : and yet I say unto you, That even Solomon in all his glory was not arrayed like one of these."

To a close observer every part of the flower has a singular interest and beauty, and is a fascinating study. We said of stamens that the anther or pollen box was generally oblong, yellow, and opening in the centre. This describes the most common or typal form of anther, but in truth, anthers are almost as varied as blossoms in form and method of opening, or dehiscence; the color also varies from yellow to orange, brown, dark purple, brownish red, and almost black. There are flowers where the corollas are wanting or inconspicuous, and the long tassel of brightly-colored stamens constitutes the chief claim to beauty. Sometimes the stamen is like a little leaf, with dots of pollen sacs scattered over its surface, as in the cycads. The long filament may hold the anther balanced at the centre on its slender tip, the whole appearing so frail that even a breath might destroy it, yet holding its own against wind and storm until the pollen is ripened and some insect pollen-bearer comes brushing heavily against it, in its velvet coat. Again, the erect filament resembles the shaft of an arrow holding on its top a two-lobed anther like an arrow-head. These lobes when ripe open the entire length of the side, releasing the pollen. There are flowers having very long stamen filaments, which are doubled over in a loop upon themselves, as one might bend a strip of whalebone, holding the ends together. When the anthers are ripened these long filaments suddenly snap out straight, as if one released one end of the strip of whalebone. The result of this sudden snap of the filament is to scatter the ripened pollen abroad or shower with it the insect whose intrusive touch released it.

It has been noted that all parts of the flower are really modified leaves; thus the stamen is a pollen-bearing leaf, and as the pollen is of immense interest to the plant this stamen is a leaf in a high state of usefulness. Therefore when stamens begin to revert to the leaf form, by changing themselves into petals, this process is called by science " degeneration." Perhaps no flower will give us as interesting instances of the stamen harking back to the petal shape, as the white water lily. This flower has a large number of brilliant golden stamens with short filaments and large anthers, seated in the snowy fragrant heart of the corolla, and gently rocked in it by the water as in a beautiful boat. One might think the stamens

were well enough off, and that discontent could not
enter so rare a palace of delight. However, close
examination will, perhaps, show us a series of stamens
leaving their high estate, their golden usefulness, to
become petals; not only so, but when degeneration
begins there is no knowing where it will end, and
back of the white petals these stamens on the down
grade go, to turn into the green pink-tinted sepals.
Here is a calyx sepal, one edge curled over, thicker
than the others, yellowish still; what is this? Surely
here is a stamen degenerate, the short filament and
the golden anther expanded and changed broadly
and coarsely into a sepal nearly like the other sepals.
Now search the white corolla segments, and here is
one along one margin of which we plainly trace half
of a filament and half of an anther, the other halves
having bleached and widened into a petal; here is a
narrower petal showing the stamen formation more
fully; here is a stamen with its filament widened
and whitened, and its anther paled and shrunken,
well on its way to petaldom; and here are others less
and less modified, until we have the whole series
back to normal. Well, and what of that? Perhaps
in the very long ago a water lily was but a receptacle
full of stamens and pistil, and a scale or two, to prevent the intrusion of water; then these stamens took
to changing and changing, becoming at last the

numerous petals, snowy white, to attract insect guests; fragrant, also, to attract them; green-sepaled to preserve the life and beauty of the flower from the water until the ovules were fertilized, and this so-called process of degeneration has given us the large, pure, sweet white water lily. Was it, all things considered, really a degeneration?

Roses show us this change of stamens to petals. The normal rose has five large petals, braced by five strong, small, green sepals united over the ovary, and in the centre of this beauteous whorl numerous stamens. Now among the double roses we often find bent petals, with a more or less well-defined stamen upon one edge. Such developments are not freaks or sports of nature; they are the stages of a process. The pollen is so fine a dust that its several grains have no particular definition to the naked eye—it is merely yellow, brown, black, reddish, or purple dust, no more. Put the ripened stamen under a powerful microscope, and we are whirled off as by enchantment into fairy-land; never had royal princess such necklace or tiara, never had queen such crown, never wore any hand such jewels as are here displayed. What do we see here? Woven and chiseled gold; pearls, diamonds, rubies, all instinct with life, flashing, glowing, palpitating, like water in a vase; here are the crown jewels, treasured up in the sacred centre of the flower, jewels which mortals seldom see, or of which few even suspect the existence.

Crown jewels have always been temptations to daring robbers, and the object of many deep-laid plans of theft; the plant's crown jewels are subject to the same dangers. How many predatory ants and roving beetles or wasps make daring raids upon the flower, and plunder the pollen, devouring it, carrying it off in their intrusive coats to waste it upon leaves and stalks as they ramble on their way. The bees and the butterflies are not freebooters; the plant puts her jewel treasures in their keeping to pledge for large returns, or to secure strong allies and increase of state.

In the month of May flowers crowd upon us in numbers so great that we are at a loss for time to study them. Even if April has been cold the matchless arbutus has found time to bloom above last year's protecting leaves, and has passed away, leaving only a memory of its fragrance and rosy beauty. The dandelions—jolly, popular, child-beloved gold of the spring—have bloomed, and in May the grass is covered with their delicate clocks; we still in early May find the oxalis almost making a carpet for pasture lands or sunny hillsides. When the oxalis grows in damp shade its flowers and leaves are larger and of a deeper color, but the blossoms are fewer.

The leaf of the oxalis is three-divided, like the coarser leaf of the clover. Some hold the tradition that it was the oxalis and not the shamrock leaf which good St. Patrick took to prove the possibility of Trinity—one in three. Some think that really the oxalis and not the clover was the shamrock of the ancient Irish.

In the southern parts of Europe the oxalis blooms very early, soon after Easter, and is called the hallelujah flower, from the anthems sung in churches at that time.

The corolla of the oxalis has five petals, the claws forming a tube, the shorter calyx has five sepals; each flower grows solitary on a stalk. The stamens are ten, and the single pistil has five styles, showing that here are really five pistils united in growth. The golden oxalis is common in some parts of the country, the rose-colored oxalis in others; the rose-colored oxalis has almost white individuals, and the blossoms are usually larger than those of the yellow variety. As the season advances the plants continue to bloom, but the flower is constantly smaller. The oxalis is one of the flowers which sleeps at night, rolling up its corolla and folding its leaves.

May brings us an abundance of wild violets; the blue violets, and the beautiful tri-colored pansies come in April, but the blue violets linger, growing

7

larger and richer, while their cousins, the dainty white, and the branching yellow violets, appear in the cool, damp woods. These wild violets are scentless, except for the spicy " woods odor" that seems to hang about all wild flowers.

A much humbler flower than the violet greets us on the roadsides—the bright yellow cinquefoil, its vine, leaves, and blossom bearing resemblance to the strawberry, so that country people call them " yellow-flowered strawberries." Common as the cinquefoil is, it belongs to a noble, even royal family among flowers—the rose; it is "a poor cousin " of the garden's queen.

CHAPTER VI

SOLOMON'S RIVALS

JUNE

" The purple heath and golden broom
  On many mountains catch the gale ;
O'er lawns the lily sheds perfume,
  The violet in the vale."

JUNE is the month of flowers, as May is of leaves ;
it is also the month of richest and most varied colors.
The leaves are in their prime, and the flowers are in
their greatest luxuriance. Green is the chief color
in the plant world. It is so constantly the color of
leaves that " green as a leaf " and " leaf-green " are
common expressions.

Many plants have parti-colored foliage, but in comparison with the whole they are few, and when we
ask what is the usual color of leaves, stems, bracts,
tendrils, or calyx, " green " is a fair answer.

Next to green, yellow is the most common color.
This is the chief color of stamens and pistils ; it is
also the chief color of spring flowers, though white
blossoms also abound. When the warmth and glow

of the sunshine returns to us, many flowers seem to assume the livery of the sun ; flecks and streaks of sunshine gleam at us from hundreds of nooks and corners. The bluebird and the jay come to us, decked in the clear blue of the spring sky, but the spring flowers outbid them a hundred to one in choosing the " class color " of the season.

The very thought of spring is associated with dandelions, buttercups, hawksweed, mustard, cinquefoil, primroses, cowslips, marsh marigolds, adder's-tongue, and a hundred other yellow beauties, from the inconspicuous parsnip and wild radish to the sunny splendor of the meadow lily and lady's-slipper.

In Alaska the flowers are nearly all yellow or white ; blue and pink blossoms are exceptions.

After yellow, white is the most frequent color, and we recall a succession of blossoms, from little white chick-weed through star-flower and trillium, bunchberry and bell-wort, up to the great, white, fragrant lilies. Next in order of abundance comes blue, then pink, purple, red, and, least seen of all, that rich scarlet that graces the cardinal flower and the salvia—

"In emerald tufts, flowers purple, blue, and white,
Like sapphire, pearl, and rich embroidery."

Autumn is the season of the most brilliant colors,

SUMMER NOON

for then come the changed forest leaves and the gorgeous hosts of dahlias, asters, and chrysanthemums.

Insects seem to have preferences for special colors. Sir John Lubbock proved, to his own satisfaction at least, that bees prefer blue, flies dull yellow or flesh-color, moths white or lemon-yellow.

The color in plants is not confined to leaves and blossoms. It is spread in stems, wood, bracts. We have only to look into a seedsman's window to see green, yellow, brown, white, pink, red, purple, lavished upon seeds. Possibly the next window will be that of a fruit store, and what can outvie the gold of oranges and lemons, the pink of peaches, the purple of grapes and plums, the mingled colors of melons and apples, the deep glow of pomegranates, the vivid ruby of currants and cherries?

Whoever called roots "dull brown things" has not looked well at beets, carrots, turnips, the snowy white, the orange, and red roots laid bare in forest or garden by a little digging.

In a study of plant colors let us begin with green. The green matter in the leaf is a soft pulp, composed of cells, and is called chlorophyl, or leaf-green; leaf-green let us call it. This is a very important substance, not only in preparing plant-food, but in purifying and oxygenating the atmosphere.

Much as leaf-green has been studied, no one can tell exactly of what it is made or how it performs its work. We only know that the work is very well done; food and fresh air come freely to us from these busy cells, the workshop of the leaf-green.

A pale yellow pulp called etiolin is supposed to precede the leaf-green. This sickly yellow is the tint of most leaves in the bud, of the first unfolding of cotyledons, and of plants grown in darkness. If green plants are put into a dark, chilly place, their leaf-green returns to the dull etiolin.

The change made in such leaves by exposure to light and warmth is very rapid. Take a potato vine; let it sprout in a cellar. It reaches out toward the place where there is light, even if that comes but from some little chink. Toward that light the pale vine grows, no matter how it has to twist and turn to do so. Let in stronger light somewhere, and very soon it has changed its direction.

A vine thus grown in darkness will be nearly white, and also soft and watery. Place it in the sunshine; in a few hours it will be green; in a few days dark green and firm of substance. A very few hours will change the etiolin to full green, and the rapidity of the change is in proportion to the sunlight given.

From simple experiments we see that the pale etiolin, which is formed chiefly from starch, is the predecessor of the leaf-green, and changes to leaf-green by light and warmth. To maintain this leaf-green the plant must have plenty of carbonic acid gas, and but little oxygen.

Here is another curious fact: When leaf-green is taken from the leaf and compressed and freed as far as possible from other substances by chemical treatment, if it is made into a thick layer and looked at upon the surface, it appears of a deep lake red. Make the layer thin and look through it, and it is of a rich green.

A plant naturally having green leaves cannot prepare proper food stuff without leaf-green. If it is robbed of this chlorophyl by darkness and cold, no food will be prepared in the leaves, and the plant will soon starve to death under the insufficient feeding of the root.

Some ferns have been known to take on leaf-green if they had plenty of warmth while in darkness. This suggests that the change from etiolin to chlorophyl is in part a cooking process. In plants where leaves and stems are by nature gayly colored, the work of the leaf-green is performed by the soft, brightly-colored pulp which takes its place.

When plants become parasitic and suck out of other plants the leaf-prepared plant-food, they cease to use their own leaves, and these leaves turn into colored bracts. The beech-drops, or Indian-pipes, are parasitic on roots, and become entirely waxen white or pale yellow. The dodder changes its green stem to red, and divests itself of leaves entirely. The mistletoe keeps its green leaves because it is not entirely parasitic, but continues to manufacture plant-food for itself.*

Trees that shed their leaves in autumn are called deciduous. Such trees, before losing their foliage, are very gorgeous, the green of the leaves changing to yellow, purple, scarlet, orange, bronze, brown. This change is produced chiefly by the withdrawing of green color, because the clogged cells are no longer taking in carbon or discharging oxygen. The excess of oxygen now changes the failing leaf-green itself to some gay color, or aids other colors which have lain hidden to gain strength and assert themselves, appearing in glory in place of the vanished chlorophyl.

Along with leaf-green in the plant other colors lie, even in stems and leaves, ready to appear whenever room is made for them by the departing green, and strength is added to them by oxygen. In the vari-colored foliage, for some reason peculiar to the particular plant, the chlorophyl politely admits a proportion of other colors to share its reign.

---

* This subject will be discussed in a future chapter.

This brings us to the consideration of chromule, which is the name given in general to all colors except green in the plant. These colors are more or less present always. In early spring the young leaves of the oak are pink and bronze; those of the birch are purple; other trees show in their fresh leafage scarlet and yellow, almost as gay as autumn dyes.

The petals of the flower, however, assert their claim to the chromule, and appear in the most brilliant colors, which are stored up in the tissues. All the hues of the rainbow, and all the shades produced by the mingling of these, belong to corollas.

> "The daisy, primrose, violet darkly blue,
> And polyanthus of unnumbered dyes."

Here we stand before some of nature's deepest mysteries and most marvelous chemistry. Out of the same soil—light, air, moisture, heat—the cardinal flower draws its flaming scarlet, and the violet its heavenly blue; the lily secures its cup of pearl in the same laboratory from apparently the same conditions that gave the golden rod its yellow plume. Look at a tiger lily; it is orange, spotted with brown; here is a rose streaked red, white, pink; here is a morning-glory painted in white, rose, blue, purple; behold this painted cup, the sharpest contrasts of green, yellow, scarlet—yet not one shade in all these intrudes upon another.   Bryant says—

"Scarlet tufts
Are glowing in the green like flakes of fire;
The wanderers of the prairie know them well,
And call the brilliant flower the painted cup."

Oddly enough, in this flower the vivid red is in leaves or bracts, clustered among the pale yellow bloom.

Many children have used the petals of flowers as paints, finding in them abundant deeply-dyed juice. The three blue petals of the spider-plant, the pink of rose petals, the yellow of the orange or tiger lily, the abundant crimson of poke or alder berries, have painted many a child's work of pictorial art.

From this color in plants, commerce obtains valuable and lasting dyes—as from the indigo, oak, madder, saffron.

Although light seems to have such influence on the production of color in plants, we find many plants deeply green, or gaudily colored, that grow where there is little or no light.   Sea-weeds of intense green, or painted as gayly as parrots, come from depths under water where the light must have been very dim.

The most vivid colors are often found in the

mold in jars of preserves that have been kept entirely in the dark. This mold is a vegetable
growth.

In the spring one may notice early in the morning
a bed of chickory in bloom; it is of a clear, exquisite
blue; by ten o'clock the blue will be very pale, by
twelve the flowers are white, by one they are all
folded up, to open next day as richly blue as ever.
The sun plays such tricks on blue cotton cloth.
Where it is exposed to the sun the blue vanishes, and
when the cloth has been put away in darkness the
fled color returns. Other flowers besides chickory
grow pale with excess of light, just as some grow
pale from darkness. The study of color in the
plant world affords opportunity for interesting experiments.

"Solomon in all his glory was not arrayed like one
of these!" How this rings in our minds when we see
the Summer Color Exposition! Yet June cannot be
to us a mere color study in general, for her prodi-
gality of individual blossoms calls us to a closer investigation of particular or peculiar plants.

Bryant calls June "flowery June." Coleridge
calls the month "leafy June"—it might also be
called fragrant June, for it seems the natal month of
the most fragrant flowers. One may notice sometimes in Scotland, such a rich, almost overwhelming

fragrance on a ripple of wind that one will stop to consider its origin. The source is not far to seek; it is a bean-field in full blossom, and it recalls what Isaac said of his favorite son, Esau, "The smell of my son is as the smell of a field that God hath blessed." No doubt that genial freebooter had taken toll of many a passing caravan, bearing myrrh, aloes, balm, and cassia into Egypt, and therewith had made his goodly garments fragrant.

Another very common and fragrant family of June blooming plants is the trefoil or clover family. A field of the common red clover in full bloom will rival in richness of perfume the famous Scotch bean-field. The clover field is a blaze of beauty—full, round, rosy heads spread under the sunshine a cloth not of gold, but of a purplish pink, strong, healthy plants these, full of suggestions of vigor. Across this field boom thousands of great humble bees, and here we are reminded that the humble bee is the especial partner of the red clover, carrying its pollen and paid by the honey from its deep cups. The red clover is not a native flower in America. It was introduced here, as in Australia, from England, but seems to have taken a special hold on the hearts of the people, and upon the soil as well.

"What airs outblown from ferny dells
Of clover bloom and sweet-briar smells,"

says Whittier, the poet of the home and of country-
side. Another sings :

> " Crimson clover I discover
> By the garden gate."

Each head of red clover is composed of hundreds
of little florets, tube-shaped, each with its own tiny
calyx, each with its nectary honey full and deeply
hidden, each pouring its portion of perfume upon
the warm June air.

Another clover becoming naturalized, a stranger
from afar, is the crimson or Hungarian clover. The
head of blossoms is not round, but long, shaped like
the first joint of a forefinger. The color is a rich,
vivid crimson, or blood red, from which it receives
its name, Trifolium incarnatum. This is the clover
so provided with tough hairs that it has proved
injurious to cattle.

An entire contrast to the large and showy Hungarian clover is the modest, low-growing, dainty,
white clover, its fragrance equally delicious, but
more subtle and delicate. The white clover has
a short, simple stem, its leaves are much smaller
than those of the red clover, and the plant hugs the
ground, having a running habit. It is so tenacious
of existence and such a close grower that where it
once possesses the soil it is capable of crowding out

the most noxious weeds, even the much detested plantain, terror of the lovers of handsome lawns and well-kept roadsides. Among these white clover will be found heads of a larger growth, more rounded and of a pure flesh tint. The head of white clover is somewhat flattened, is loose, and has a greenish tint under its whiteness, because the little green calyx of each tube is partly seen.

Frequent on dry, sandy roadsides, especially by railroad beds, is the little yellow hop clover, often not recognized by the casual observer as a clover at all. The poor soil which it most affects checks its growth, and it is a slim, stunted herb, its foliage pale green, with leaflets more slender than those of other trefoils. When the seed has ripened, the small, long, downy heads look oddly like the catkins of the pussy willows.

Leaving the roadsides and the clover bloom and entering some long undisturbed " wood lot," or passing near thick undergrowth beside some little brook, a rich, special fragrance greets us, more delicious than any spicy waft from Araby the blest. It is lavishly poured upon the air by the inconspicuous green blossoms of the wild grape; once we meet a breeze heavy with this exquisite odor it is never forgotten, and the sensitive nostril is likely to try all other perfumes by that one unattainable standard.

Lines previously quoted link the clover and the sweet-briar; here we consider that June marshals in the troops of roses, among the choicest of its fragrant bands—" sweet as a rose " has passed into a proverb. The lilies also are June flowers; the golden lily, the stately white-lily, each one pouring from its chalice a wealth of sweetness, each one fair enough to be the long-sought Holy Grail. With the roses and lilies come the honeysuckle tribe, and all together vindicate the claim of June as the especial month of fragrance.

There are some of the children of June which find their " excuse for being " not at all in their fragrance, but in their singular beauty, and of these the laurel stands pre-eminent. No one who has climbed the hills near Gettysburg when the laurel is in full bloom, flushing the hills from base to crown with a deep rosy glow, can fail to accord the mountain laurel the palm of beauty. Its pink tint is the most exquisite tone of pink that nature ever produced; its buds, each one like a choicely wrought jewel-box, with a deeply-fluted lid, are the most beautiful buds that could be contrived—the open blossom a more delicate pink than the buds, and given a lace-like appearance by the long filaments of the ten stamens, is indeed a matchless flower, while nothing in the marvelous cross fertilization of orchids can surpass the ingenious contrivance of these flowers to secure also their cross fertilization. The corolla is wheel-shaped, with five lobes; each of these lobes has two depressions; in each of these ten depressions of the corolla lies one of the ten stamens, its long filament recurved and its anther, the true jewel of this jewel-box, carefully lodged in the tiny pink niche which exactly fits it. So firmly are these anthers held back that the filament is arched in a little bow above the surface of the petals. When the anthers have ripened their pollen they are large for their resting-place, and their hold upon it is loosened, but they retain their position from habit, until some insect, lured by the pink chalice, alights upon it, or passing bee or butterfly brushes it with fleet wing. Then, in a second, all these little resorts snap back, up springs each stamen with violence, the anthers are jerked from their niches, the pollen flies out in such a golden shower as fell over Danae, the destiny of the blossom is accomplished, its stigma has received from the body of its insect visitor the pollen of some other laurel bush, and this laurel has sent its own contingent to blossoms far away. Happy, indeed, was the Swiss Peter Kalm, the assistant of Linnæus, to have his name immortalized by this lovely plant, the mountain, or wide-leaved Kalmia.

Near the seacoast we find a tiny pattern of this

splendid laurel in a shrub scarcely a foot high, bearing leaves and flowers dwarfed in proportion, the blossoms of the same shape, but less lovely coloring than those of the mountain laurel—the sheep-laurel, supposed to be very poisonous to flocks and herds. When the mountain laurel grows in the partial shade its flowers are of a paler pink than when they receive the full largess of the sun. When a stray bush retreats fairly into the cool shadows of thick woods, the flowers are fewer, larger, and of a pearly white.

A near cousin of the laurel is the pink azalea, with large, almost white and richly fragrant flowers. This shrub frequents wet lands, near which grow large trees that yield them shade from too powerful sunshine. The azalea, or wild swamp honeysuckle, blooms earlier than the laurel, and a single bush crowded with bloom will send its fragrance abroad for a mile or more. The peculiarity of this odor is that at a distance, borne on the wind, it is delicious; near by there is a certain deathly heaviness about it. Emerson finds in a cousin of the azalea and the laurel "a rival of the rose." That rhodora was the famous plant which taught the poet that "beauty is its own excuse for being."

The elder sister of this beauteous rhodora, the greater rhododendron, is, like the laurel, a mountain lover. The flower stalks of this plant are particu-

8

larly viscous or sticky, a contrivance to keep crawl-
ing creatures of the ant kind away from its stamens.
The single pistil is much longer than the stamens,
and is somewhat recurved. The corolla is slightly
bell-shaped, and when a bee, dusted with pollen from
a flower previously visited enters the corolla after
honey, the pistil cannot fail to brush his coat, and
collect the pollen upon its sticky style. Then the
stamens are reached, and a fresh load of pollen dusts
the bee's coat; this pollen from its own flower the
style does not gather, for its sticky surface is already
covered with yellow powder.

CHAPTER VII

PLANT PARTNERSHIPS

JULY

"My trees are full of songs, and flowers, and fruit,
Their branches spread a city to the air."

"While the bee with cowslip bloom was wrestling."

THE production of seed is the chief object of plant
life. Upon this depends the continuance of the vegetable world, and therefore of all animal existence.
From the elephant to the mouse, from the whale to
the minnow, from the eagle to the tomtit, life is con-
ditioned upon the constant return of "the herb-
bearing fruit whose seed is in itself."

In every minute particular the flower is con-
structed to assure the production of sound seed.
The first form of this seed is the tiny ovule in the
germ. Ovules cannot grow into seeds, unless they
are reached and fertilized by the pollen, which must
arrive at them by the way of the stigma.

We owe much to those careful students who have,
by close observation and careful experiments, found

115

out for us the facts that concern the fertilization of the ovules. What are these facts?

I. The pollen of the stamen must reach the ovules through the stigma of the pistil.

II. The most vigorous and perfect seed is secured when the pollen from one flower is conveyed to the pistil of some other flower, rather than to its own pistils. The ovules become better seed if they are fertilized from blossoms borne on a different root.

III. Nature has many curious contrivances to prevent pollen from reaching pistils contained in the same corolla as the anthers from which it came. For instance, the anthers and stigma in some flowers are so situated that the pollen cannot pass from one to another; in many other cases the stamens and pistils do not ripen in the same flower at the same time.

IV. The most wonderful arrangements have been made to secure the conveyance of pollen to distant blossoms.

V. Unless pollen is conveyed to plants of the same kind as that producing the pollen, seed will not be fertilized. Lily pollen cannot fertilize the ovules of roses; buttercup pollen is wasted on dandelions.

VI. The animal world, which is maintained by the vegetable world, must provide the carriers for a large part of this pollen, for the flowers are root-fast.

VII. This carrying is not to depend upon the good-will or intention of the carriers, but must be a necessary happening of their own existences.

Let us see how all this is secured.

The pollen of flowers is a most fine, delicate dust. It must be conveyed without injury in the most accurate manner. Many flowers are exceedingly high up, as on climbing vines, or growing on tree-tops, peaks, or house-tops. Many other plants are very low down, lying close to the ground, as the bluets, chickweed, arbutus, partridge-berry, and others. A large number of plants are in positions inaccessible to man or the larger animals.

Man excepted, the larger animals seem generally to have a destructive mission to plants, devouring, breaking, or trampling down. Men themselves are often ruthless destroyers of beautiful plants, and seem generally to care for and conserve only what concerns human convenience.

Here, then, we have the problem of plants fixed in their places, needing carriers for their pollen to distant plants of their own kind, at the exact period of maturity. The carriers must be able to go high or low, into all manner of difficult localities; they must be delicately made, so that they will not injure the plants which they visit, capable of carrying the frail pollen grains unharmed, and they must have some object of their own in these visitations, which shall infallibly secure their doing of the work required. Finally, let us remember that the pollen of flowers is but seldom spread where it is easy to secure it. The buttercup lavishly expends a golden saucer of pollen; the lily has a wide-open door, near which hang the anthers, like so many ready bells. On the other hand, how long and narrow are the throats of morning-glories and honeysuckles; how tiny are the tubes of mint, thyme, and clover; how fast-closed is the mouth of the snapdragon; how narrow the fox-glove's throat. Pollen-carriers must be able to secure the dust so jealously kept, and must be afforded a reward for their trouble.

What form of animal life meets all these conditions? But one. The insect. The insect is generally light and delicate in structure, active, winged; its life is conterminous with that of the flowers; they are spring and summer guests. The slender shape and the long, slim mouth organs of the insect can penetrate and gently force open flower tubes and the fast-shut lips of corollas; the velvet coats and fine waving antennæ will receive and carry uninjured the precious dust, and the insect habit of constant roaming from bloom to bloom assures the accomplishment of its important errand.

Not all insects, but a few widely distributed

families, are chosen partners of the flowers ; these are the various tribes of bees, moths, and butterflies, with some help from a few others.

" Nothing for nothing " seems to be a law in nature. What does the flower offer to the insect for its services as pollen distributor ? Honey, which is the chief food of flying insects, also wax, and pollen for its private use at home. The miller, we know, takes toll from the flour he grinds.

To secure insect visitants the flower provides in its nectaries honey; almost all flowers secrete some dainty juices. As shopkeepers set up signs to inform the public of their wares, so the flowers hang forth signs ; these are the brilliant corollas, or parts highly colored which do the office of corollas.

The pea, which hides stamens and pistils under a close cap, spreads above them brilliantly painted banners, meaning " honey on draft," and the honey is hidden way back in the stem end of the keel. When the bee lights on the wings of the corolla his weight pulls down the cap and releases the stamens ; then one stamen springs back, leaving room for the bee to seek the honey at the bottom of the stamens, but as he does so he is covered upon his breast with pollen.

The compositæ usually detail a ring of wider-petaled florets upon the outer edge of the compound flowers, simply as a splendid sign of the whereabouts of honey for insects coursing through the air.

A writer on flowers says the compositæ owe their wide range to their "co-operative system." If we look at a daisy we see that its centre comprises a whole mass of little yellow bells, each of which consists of corolla, stamens, and pistil. The insect which alights on the head can take his fill in a leisurely way without moving from his standing place. Meanwhile he is proving a good ally to the plant by fertilizing one or another of its numerous ovaries with pollen gathered on some other daisy. Each tiny floret alone would be too inconspicuous to attract attention from the passing bee; but union is strength for the daisy as well as for the State.

Another bid for visits is made by perfume, which attracts insects as being generally associated with honey. Many flowers have inconspicuous corollas, or are hidden under foliage, or so placed as to risk being neglected; these call attention by fragrance, as the mignonette, the violet, or arbutus. Others, as lilies, have large and attractive corollas, yet add perfume to size and color, to insure the securing of insect attention.

Plants which depend upon moths, or any night-flying insects, have usually strong perfume and pale color, as white or light lemon color, which can easily be seen in twilight. The odor attracts the insect in its direction, and on a nearer approach the flower is seen.

Most flowers have peculiarly bright streaks, spots, or other markings, in the direction of the honey, and the honey is placed at the bottom of the stamens; thus the insect is attracted just where he should go. The tiger lily has its startling black spots, the arum its lines of red and green, the morning-glory its vivid stripes, the jonquil its ruffled bi-colored crown, the beauty-of-the-night its rich purple centre.

When the pollen is ripe for carrying, all the parts of the flower are at their best; the perfume is the strongest, the coloring is the brightest, the nectar most abundant.

Now does some one say, " If the bee is wrestling with the bloom, why will he not rub the pollen of the anthers into the pistil of the same flower? Thus the very end to be avoided would be directly secured." Why not, indeed? O, wonderful plan of the flower! Here is the beautiful explanation. The pollen cannot adhere to an unripe stigma—only the mature stigma provides the sticky secretion that will retain the pollen grain. In nearly all flowers the stamens and pistils do not ripen simultaneously, but the one a little before the other. In some flowers the pistils ripen first, but in most the anthers ripen and have disposed of all their pollen before the stigmas have come to maturity.

A bee comes to a foxglove; he is well powdered with yellow pollen, which was shed upon his velvet coat while he sipped honey in some other foxglove. When he enters this fresh flower the pistils are ripe and their viscid stigmas are all ready to catch the pollen from his coat while he is eating, and hold it fast when he goes his way, but the stamens of this present flower are not ripe, and give him no burden. To-morrow he or some other bee will return, and the ripened anthers will send their store to other flowers. If all flowers of a kind opened on the same day this distribution could not be accomplished, but there is always a succession of bloom for days, or even weeks.

Here is another query: As the pollen cannot fertilize seeds of another kind than its own, and these insect rangers ramble where they will, why does it not happen that in nine cases out of ten they go from flower of one kind to entirely another kind, as from asters to golden rod, from daisies to roses, from buttercup to clover?

The insects are not such reckless expressmen as that. They carry the pollen parcels to their proper places, because—here is a wonder of wonders—because their fixed habit is to go from one flower to

PLANTS AND THEIR PARTNERS

another flower of that same kind, not to one of another family. They seem to have a rooted aversion to mixing their food or their honey cargo. From clover to clover, from pea to pea, from gladiolus to gladiolus, that is the bees' style of doing business.

On these hot July days, when the sun draws out the richest fragrance and lights up the most brilliant colors, watch the bees and the butterflies. The bee seeks clover on one trip, mignonette on another, lilies on a third. The butterflies have no hive returnings to mark their work, but you can count their visits, a dozen or more to flowers of one kind before they investigate the sweets of flowers of some other kind.

Corollas are generally so fashioned that the insect does not receive the pollen promiscuously over his body, but brushes it off the anthers at exactly that place—proboscis, head, breast, shoulders—where it will be exactly in the way of the stigmas of the next flower of the same kind visited.

Volumes have been written about the manner in which the pollen grains clasp the antennæ or proboscis, or stick to the top of the head or the shoulders of the insect guests.

In spite of all these precautions an immense amount of pollen is lost in one way or another, especially in wind-fertilized flowers. This loss can be sustained without injury to the plant-world, because the production of pollen grains is so enormous. Many flowers produce three or four million grains of pollen. Dandelions are saving of pollen, but even they yield about half a million grains to each flower. The Scotch fir produces so much pollen that it blows from the tree like a cloud of smoke, or pale yellow mist.

There are some flowers where the pollen is sent to the pistil in the same corolla. This is called self-fertilization. Self-fertilization is the exception, not the rule. Many flowers cannot produce any seed at all without the aid of some especial insects. Thus the red clover can only yield seed if visited by humble bees, because these only have a feeding-tube of the proper size and shape to bring them into contact with the pollen.

All flowers do not have bright corollas, nectar, and perfume to attract and reward their helpers in the seed business. What of them? What of those tiny, fringe-like, delicately-colored stamens and pistils held in a small, pale, scentless scale, which make up the flower of the grass? What will the grass do while

"The broom's betrothed to the bee?"

Many tree blossoms are in the same case. What has the elm or the beech to reward or lure the bee?

These plants do not require the visits of insects. They have secured the services of a much more potent ally—the summer wind.

Look at a field of grass in its brief bloom. The stamens and pistils hang from the clasping scales. Along the field sweeps the gentle wind, swaying the graceful stems until they ripple like the sea. As they bow and rise freely upon the wind's soft wings, the pollen floats away, blown abroad the field and neighboring fields, and so reaches millions of waiting pistils. In this same manner also are the germs on many trees fertilized. The pollen of the thorn and poplar will float for miles to reach some distant pistillate blossoms awaiting it.

Wherever we find inflorescence reduced to its lowest terms, no gay petals, no sweet perfume, no choice nectar, merely stamens and pistils held in a scale or two, as in a loosely-clasping hand, we know by these signs that the wind is the partner challenged to do the outside business of the firm. For many years even earnest students of botany supposed that the especial intention of nature was that the pollen of stamens should fall upon and fertilize the pistils of the same flower; and so deeply rooted was this opinion that very evident facts pointing to a different conclusion were overlooked. Mrs. Lincoln's " Botany for Beginners " was one of the joys

of young lovers of plants. years ago. In regard to
stamens and pistils the good author discoursed somewhat on this wise: " It is of immense importance
that the pollen of each flower should reach the pistils
of that flower. and it is well to notice how this object
is secured. In flowers where the pistil is the longer
the blossom hangs downward. so that when the
anthers are ripe the pollen will be shed upon the
stigma. Where the pistil is shorter than the stamens
the flower is erect. so that the pollen will still drop
from the stamens directly upon the pistil. For an
example of flowers with pistils longer than the stamens examine the lily." This was the substance if
not the letter of the work, and one accordingly " examined the lily " and was filled with admiration of this
beautiful contrivance for securing self-fertilization.
No one at that date seemed to have thought of
noticing several very evident facts which disproved
their whole theory of lily fertilization. First. the
anthers and stigma do not ripen at the same time;
the ripening of anthers is shown by the opening
of the anther-case and the dissemination of pollen; the ripening of the stigma is shown by a thin
glue or sticky juice. which exudes from its surface
and causes the pollen to adhere and germinate.
When the stamens of the lily are ready the pistil is
not. Second, the sticky surface of the stigma is the

top. Now, as the lily bell hangs downward, the under side only of the stigma—in this case merely the expanded crest of the style—is exposed to the anthers bending above, and if the pollen shower were ready it would almost inevitably fall upon the ground, and be very little likely to reach the stigma. As an example of erect flowers, where the pistils were short, the book gave the rose, and again ignored objections of equal force. All this shows how important is an unbiased examination of all the facts in a case, and the being willing to study seriously and experiment carefully, even in instances where we think we know. Such work secured the vast amount of important facts with which Darwin dowered his age.

The case of wind-carried pollen for trees where one tree bore only staminate and some other pistillate flowers, was earlier understood. The "Beginners" book gave an instance of two poplars, forty miles apart, one with pistillate the other with staminate blossoms. Of course no seed was matured until the growth of the two had lifted them above intervening obstacles, when the wind finally bore the pollen to the waiting pistils.

Hugh Miller relates that when he was a lad some girls told him of a large thorn tree which never bore fruit. Miller at once saw that it was a tree having only pistillate blossoms. He told the maids that it was enchanted, but that he could release the spell. He said some mummery over the bush, and told the admiring listeners to come in the proper season and they would find haws. Then, at night, he brought secretly to the pistillate bush a large branch of staminate flowers and brushed it thoroughly over it. His trick—or experiment—succeeded, the flowers were fertilized, and, to the amazement of the girls, the tree bore fruit.

Of all plants the orchids are the most remarkable in expedients for fertilization. Since Darwin opened the eyes of the world to their wonderful structure, there is a romance about the very name orchid, and we feel that it must be synonymous with beauty, remembering those marvels of the florist's windows, where all splendors of flower form and color seem to meet. Truth is, there are some very inconspicuous orchids, as the "green orchis" of our bogs, with small, unattractive spikes of dull green flowers. In general, when we say "orchid," our imaginations hie away to the exotic parasites for which fabulous sums have been paid. Our own woods and swamps still show numerous beautiful orchidean varieties, which are in a fair way to be destroyed by ruthless tearers-up of entire plants. Nearly every one has seen the great purple lady's-slipper, or moccasin flower. Between two large downy leaves which have protected the growing bud rises the simple scape, bearing on its top a solitary pink-purple flower. There are three green spreading arched sepals, and under these two narrow wrinkled side petals of a greenish tint, looking like sepals; from between these depends a very large pink petal, inflated and united at its lower edge into a pouch. One botanist said if he " found a hundred growing close together, each one would be a rarity." Woods, hillsides, swamps may be the home of this moccasin flower, if it can but be let alone to mature seed; but its singular beauty provokes ruthless marauders to tear up every specimen seen. A near relative of this great purple orchis is the golden orchis, or " whippoorwill's shoe." The stem of this lovely plant will reach two feet in height, and bears broad, oval, many-nerved leaves, placed opposite almost to the top of the stem. The bright yellow flower is terminal; the three sepals are brownish, and two unite under the lip; the side petals are brown, very long, and waved, we might say curly, like tresses; the lip is a fine yellow, swelling into a pouch touched with brown lines and shadows. A smaller yellow lady's-slipper is like this, except that the color is richer, and it has what many orchids lack—perfume. These large, showy pouches of the slipper orchids attract

9

bees, who thrust in their heads after honey, and so reach the short stamens and pistils. The pistils are particularly rough and moist, and to them promptly cleaves any pollen the visitor may have collected from some adjacent flower.

The "showy orchis," whose many pink and white flowers grow in a long spike, is, like the moccasin flower, an early bloomer. The leaves are nearly opposite and sheathing, growing upon the stem; the flower has leaf-like bracts protectively placed near it, it is more widely opened than the slipper varieties, and the lower lip is white, with a wavy outline. This orchis is very much like that ornament of the English meadows which Shakespeare in "Hamlet" refers to by the name "long purples."

A small, but exquisite orchis is a native of cranberry bogs; the corolla is of a raspberry-pink tint, and the fragrance is strongly raspberry; four or five of these delicate spires of bloom will perfume a whole room, and never fail to deceive people into saying, "Why, where are the raspberries?" Near this orchis usually grows a variety closely like it, yet with distinct differences, the purple-fringed orchis. Neither exceed ten inches in height, and the bloom is fugitive, while the lady's-slippers will retain their beauty for several days.

The Northern white orchis is tall, and has white

wands of bloom, delicately fragrant; in all these varieties the placement of stamens and pistils and the method of fertilization deserves careful study.

The true "long purple" of Shakespeare, which the "showy orchis" resembles, is the purple loosestrife, an English flower become naturalized in the United States. The leaves are opposite, lance-shape, heart-lobed at base, sometimes changing to whorls of three; the flowers are a deep purple-pink, with wrinkled petals, sometimes five, sometimes six. This is an instance of a flower with stamens in "two sets," there are six long and six short stamens. There is but one pistil, and this is peculiar in the varying length of the style; there are three different lengths, which appear in different flowers. This variation in the length of filaments and style is by no means accidental, but is carefully arranged for the purposes of fertilization. For such purpose the orchids have been slowly differentiated from a simple flower of an original snow-drop pattern; and so in the flaming tassels of loosestrife has developed trimorphism, the stamens varying in length, and being especially fitted and ripened to suit the three different lengths of pistil. The loosestrife is a lover of stream borders and marsh lands, and in August flushes wide spaces with the rich color of its long, irregular spikes of bloom.

CHAPTER VIII

PLANT-FOOD AND MOTION

AUGUST

"Restless sunflower, cease to move."

"Spoke full well in language quaint and olden
One who dwelleth on the castled Rhine,
When he called the flowers all blue and golden,
Stars that in earth's firmament do shine."

IN the hot August days, when the air scarcely
stirs, the birds sit silent in their coverts, the cattle
stand under the thickest shade or knee deep in the
ponds. Only the insects seem to rejoice in the
burning rays of the sun, and gayly hover around
the splendid profusion of the flowers.

In this season we may make various studies in
plant-life. Seated upon some shaded veranda, we
have the glory of the garden spread out before us.
Or we may be on some hill, tree crowned, not far
from the sea; we find within hand reach golden-rod,
asters, milfoil, blazing-star, indigo. Looking down
the gentle slope to the level land we see black-eyed
Susan flaunting beside St. John's wort, and wild

132

snapdragon. Yonder the little brooklet slips along without a ripple, cherishing on its border loosestrife and jewel-weed. Out in the roadway, defiant of summer dust, almost in the wheel-track, the mullein lifts its dry, gray foliage and unfolds its tardy pairs of clear yellow bloom beside that exquisite flower the evening primrose, of which the harsh, dusty stem and leaves are such rude contrast to the fragrant salvers of pale gold—the blossoms of one night.

We have ample opportunity in some or all of these to study the motion, food, and some of the varied products of the plant world.

Motion? What motions have plants other than as the wind sways them? True, there is an upward motion; they "grow up," inch after inch, foot after foot, the law of growth overcoming the law of gravitation. The sap rises in the vessels by root-pressure, by capillary attraction, by the forming of a vacuum in the leaf-cells by evaporation, and so the climbing sap builds up the plant. This getting up in the world is not a trifle in plant-life any more than in human life.

Many a plant seems to have an extreme ambition to rise, and if its stem proves too weak to support any decided advancement in growth it takes measures to secure aid. It twines, bodily perhaps, around the nearest support, as do the trumpet-creeper and

honeysuckle; it modifies leaves into tendrils, as does
the sweet pea; it puts forth aerial roots at its nodes,
as does the ivy; it elongates a leaf-stem to wrap
around and around some proffered stay, as does the
clematis; or diverts a bud for such purpose, as the
grape vine.

Other plants, of a lowlier mind, creep along the
ground. The prince's-pine forms a strong, thick mat,
cleaving to every root, twig, grass-stem in its way,
striking rootlets here and there, until only a strong
hand and a firm wrench can drag it from the earth,
its mother. Cinque-foil and its cousin, strawberry,
send out runners from all sides, which root and
shoot up new plants until the whole bed is a soli-
darity, and would so remain did not the thankless
plants keep all food and moisture for themselves,
and deliver over the runners to death by starvation.

The walking-fern has a most original way of getting over the ground; it bends its slender frond and
starts a root by extending the tip of the mid-rib; so
it sets up a new plant, and is anchored fast on all
sides by its rooted frond tips, covering the ground
with a rich carpet of verdure. The variety of runners along the ground is as great as the climbers up.

All motion of the plant is a form of growth. The
plant grows by day and by night, but more by day,
as light and heat are incentives to growth.

There is still another motion, so peculiar that it seems almost to be the result of plant preference. This motion is called heliotropism—that is, sun-loving or following. The name of that fragrant flower, the heliotrope, will aid us to remember this term. The heliotrope is a persistent sun-worshiper, and received its name from its constant turning toward the sun, the Greek name of the sun being *Helios*.

Let us take the sunflower as an example of sun-following. If we examine the stem we shall see that its fibres are twisted, or swerved a little, by its constant turning from side to side, in its habit of sun-worship. It is early morning; the broad golden disc of the compound flower bends toward the east, to catch the sun's first rays. At noon the flower is held erect, fair to the zenith, eyeing the glowing sun as a caged eagle does, in intense longing. As the afternoon passes, the flower bends slowly westward, and turns the same full look on the sun at setting that it did at rising. All the sunflower family are sun-followers.

Many other plants have this habit; it belongs not only to flowers, but in some cases to leaves, roots, and stems. The leaves and roots of the onion are sun-followers. That pursuit of light which we have noted, by which stems growing in any dark place seek the light, is a form of heliotropism.

Some plants, as a part of the mistletoe, turn from the sun, avoiding it; so do the ripened flaxseeds. All pot-plants show sun-following if they are so placed that one side is turned to the light and one to the shadow. The leaves turn toward the sun, reaching out their stemlets in his direction; the shaded side of the plant seems to be really bare of leafage. Turn this bare side lightwards, and in a few days all the leaves have swayed thither; the ivy-leafed geranium is a good example of this, its turning is so prompt and persistent. To keep pot-plants symmetrical in growth the pots should be turned regularly every few days unless they have light equally on all sides.

Plants have a motion of water-following, as well as sun-following. When a plant grows where there is dryness on one side and moisture on the other, the leaves and roots will be chiefly found on the water side. Note how willows and alders bend over the water by which they grow. For long distances tree roots will strike out towards water. Water-following is more common in roots, and sun-following in flowers and stems. There are some few plants which turn from moisture, just as there are some few which turn from the sun.

Plants, especially plant roots, will travel considerable distances seeking earth in which to hide. This pursuit or bending toward earth is called geotropism, or earth-following. Thus the radicle, or root-part, always grows down into the earth, while the plumule ascends. The poppy is a pronounced sun-follower while it is in bud, and an earth-seeker in blossom and seed-bearing. Note the curvature of poppy stems.

Many leaves and stems have some especial way of growing with regard to the horizon, and nothing will induce them to change their fashion. You may turn them, tie them, even nail them fast, and they will pull away and go according to their habit. Plants expend a deal of energy in growth movements of various kinds, for nearly all plant motion is a form of growth.

Gravity, light, heat, moisture, electricity, strength of fibre, sensitiveness of surface, are all parts of the "how" and the "why" of plant motion.

You can force a leaf tip or a terminal bud to write for you the story of its motions. Insert a bristle or a cambric needle in the tip of the leaf or bud; smoke a pane of glass well and lay it where the point of the needle or bristle will just touch the smoked side. As it moves the path of motion will be traced upon the glass.

Interesting as is the study of plant motion, let us forsake it and consider for a little plant-food. The plant receives food from earth, water, air. The earth gives the plant sulphur, iron, soda, magnesia, phosphorus, and other mineral substances which we have before noted. These are all offered to the plant, held in solution in water. We are sure of this, because the food must pass into the cells, the cells have walls, and only water, with what is held in solution that is changed to a liquid state in water, · can pass through the cell walls. We know that these mineral substances are carried into the plant, changed and become a part of it, because when a plant is burned, destroying its moisture, these minerals can be taken again from its ashes.

From water the plant receives as food hydrogen and forms of ammonia. From air the plants absorb carbon (dioxide), oxygen, nitrogen, and ammonia; very much of the first, little of the second, and very little of the others.

Plants seem to be as foolish as some people, in that they will absorb from the air very poisonous elements, which destroy them. When plants grow out of doors the winds, dews, and rains free the leaves from accumulations of dust which obstruct the pores and hinder the receiving of food. In very dry and dusty seasons we notice that the plants become sickly from the stopping of the pores. Plants need clean skins as human beings do. House plants should be well washed all over now and then, to admit of their getting their proper amount of food from the air.

Certain classes of plants use a portion of animal food. We are accustomed to the idea of animals eating plants, but when we see the tables turned, and plants eating animals, that is queer indeed! The animal food of the "flesh-eating" or carnivorous plants is really the juice sucked from the bodies of insects.

The sun-dew, common in marshes, expands a little sticky, pink-green shirt-button of a leaf, on which are numerous stiff hairs. The clear drops of gum attract insects to the leaf, and they are held by the feet or wings. Their struggles cause the leaf to fold together, when the hairs pierce the body of the insect and drink up the juices. When only a dry husk remains the leaf opens and the wind shakes the shell away.

The pitcher-plant invites insects by a honey-like secretion. They fall into the liquid stored in the pitcher and are there drowned, because, owing to numerous downward-pointing hairs in the throat of the pitcher, they cannot climb back. Easy is the descent into evil! The acrid liquor in the pitcher digests the bodies of the insects, turning them into plant-food. Flies, ants, gnats, little beetles are

often caught, but bees very seldom.   Bees have their
affairs to attend to, and cannot go picnicking into
pitcher-plants.

There is a little water-weed, furnished with
numerous bladders, which were once supposed to
act as floats for it, but are really traps for tiny insects—"eel traps," Sir John Lubbock calls them.
The insects crawl in, but a ring of hairs that will
not bend outward hinders their escape, and the
plant very soon has sucked their bodies dry.

The Venus fly-trap, or dionea, has an ingenious
arrangement for catching insect food.   A part of the
leaf is furnished with a strong hinge, some honey
dew and a fringe of prickles.   When an insect alights
on this pretty device the hinge snaps the sides of the
leaf together, the prickles pierce the body of the
captive, and drink it dry.   Such are some of the
varieties of plant-food.

The consideration of what plants eat suggest to us
what peculiar products they furnish to us.   Of these
we will but briefly enumerate starch, albumen, flour;
sugar, which is very abundant in some maples, in
sorghum, and sugar canes, in beets, carrots, and
other plants; gum-arabic, gutta-percha, resin, tur-
pentine, sweet or olive oil, other oils, balsams, gums,
camphor, many dyes, many useful acids, and many
very strong poisons.   The ingenuity of man has

Pitcher
Plant

Venus's Fly Trap
(Dionæa)

Duck-weed
(magnified)

Nina G. Barlow
98.

Sundew (Drosera R)

INSECT EATERS

succeeded in increasing in plants the qualities which
are most useful to him. Cultivation increases the
amount of snowy fibre in the cotton plant, and the
amount of food-stuff in the cereals. Man, having
resolved to force the beet-root to yield him sugar,
has succeeded in converting an annual root into a
biennial, because the root grown in a single season
cannot properly store up and "cook" within it
enough of sugar to make it valuable. Cultivation
directing the strength of the plant to the especial
end of its mercantile value, repressing development
in other directions, makes almost an entire change
in some plants. The Irish potato in its natural wild
state had but small tubers, which, while nutritious,
were nearly tasteless; cultivation and the production
of new varieties by seedlings and crossings, has made
the potato the most important root in the world.
For lack of this single "root," which, as we know, is
not a true root, but a thickened underground stem,
wherein a thrifty plant stores up starch and a few
other food materials, a great people were brought
into the most terrible straits, and the "Irish
famine," occasioned by the potato blight, is written
in history. We get a commentary on the dependence of the greater upon the less when we read that
the Irish nation has never since that·famine regained
its vivacity and light-heartedness.

In the month of August our attention is drawn to this especial point of the plant world as food purveyors. The plant has been sap-feeding since February, and not only has attained its full annual development, but has laid up nearly all its stores, and is turning toward the period of rest; growth is now very slow, except in some late varieties. Fruits have gained their full size, and are now ripening; roots are also mature for gathering when it is the root or tuber that is of prime food importance; the various grain crops, except the late-planted buckwheat, are ready for housing. It is a stupendous thought that all of this vegetable material was once mineral, and that in fashion which eludes human analysis and pursuit the mineral has become vegetable in the laboratory of the plant; then being introduced by eating into the organism of the animal, will by digestion and assimilation become animal; finally returning to the soil in the process of disintegration, will once more be divided into its mineral portions, to be sent again upon its ceaseless circle of growth and decay.

This thought of these transmutations and changes outdistancing the dreams of alchemy brings by some subtle association the smallest, least-known of plants into mental view—the bacteria "microbes" which were long claimed by zoologists, but have been of late

freely relinquished to the botanists. These bacteria
are now granted to be the simplest plant forms, and
related to the algæ, or sea-weeds. When one states
that they are so small that fifteen hundred of them
could stand in line upon a pin's head, and many of
them so diaphanous that they cannot be studied
under the strongest glass until their texture is re-in-
forced by some kind of dye-stuff, we shall conclude
that here we have reached the "infinitely little."
No wonder that for ages the very existence of bacteria
was unknown, and that their vast masses when made
evident, as by the color of some of them, were supposed to be single individuals, as for instance the
myriad clusters of them which by land or water produced phosphorescent life. The commonest forms of
these minute plants are rod-shaped, spirals, spheres,
or egg-shaped; of the largest, twenty-five thousand
in a line will occupy an inch; of the smallest, double
that number. They multiply by division. One bac-
terium is a single cell; it divides, and there are two;
each of these divides, and lo, four; another division,
eight, and so on through infinite geometric progres-
sion. It is said that if let strictly alone, an unbroken
series from one would in five days fill all oceans with
a mass of bacteria a mile deep! Fortunately the
series is always interrupted; everything destroys
them; yet they seem to thrive in every circumstance

and place. Heat destroys them—yet they survive cooking; cold destroys them, yet they flourish in a block of ice, and are exuberant over a heavy freeze. These vegetable atoms disport themselves like animals; in water they swim like fishes and dive like little boys, and whirl and antic until one would fancy they must be guided by instincts or intentions. There is probably no breath of air, no fruit, no vegetable, no inch of soil free from some of the countless hosts of these bacteria. If they were necessarily destructive of human life the human race would have perished long ago. On the contrary, by grace of bacteria some of our most important food stuffs are food. Cheese and butter, for instance, would be without good flavor if the bacteria proper to their production were lacking. Also there are bacteria that make havoc with both butter and milk, so that while the one kind of "microbe" is to be encouraged the other is to be antagonized. Thus in the plant world we find organism growing upon organism, parasite upon other plant, and microbe upon parasite, and all of these with their especial uses in the economy of plant or animal life.

From small to great, from the microscopic "microbe" only lately admitted to be a plant at all, we turn to the bright succession of summer's larger children, the beautiful flowers, more precious than ever now that so many of their kindred have faded and perished, and that two months more in our Northern climates will complete their reign, and bereave us of their presence.

Along the roadsides that brilliantly blue flower, the "simpler's joy," grows abundantly. If the long spike of blossoms opened at the one time, this would be among our flower favorites, but the spike is rendered unsightly by the hard, rough seed vessels of the lower blossoms, while the florets at the top of the stalk are just opened. Another misfortune of this honest plant is, that owing to the localities of its choice the rough stem and leaves are usually loaded with dust. In spite of these drawbacks the vervain, or "simpler's joy," is a plant of song and story. The Latin poets tell us that it was a favorite decoration for altars, and was made into garlands for beasts that were to be sacrificed. No doubt the wreaths fashioned by the impulsive people of Lystra, when they were about to do homage to Paul and Barnabas as epiphanies of Jupiter and Mercury, were made of this blue simpler's joy. In the Middle Ages this plant was called "The Herb of Grace," and is said to have first bloomed upon Calvary. The name simpler's joy has no reference to idiots—"simplers" were formerly the gatherers of medicinal plants, "herb doctors," and these made

10

especial use of the vervain, as it was believed to be a specific in many diseases.

August is the gala day of the wild asters—white, blue, pink, and purple—the white being further varied by a difference in the color of the centres, some being brown and some yellow. Asters are so prolific in bloom and have such intense blue and purple coloring that they are a conspicuous feature in the August landscape. Fifty-four varieties are denizens of the Northeastern States, and of these all but twelve are blue or purple. An open hillside covered with wild asters, golden-rod, ironweed, and chicory is a splendid display of color, all made more dazzling and vivid by the blaze of the August sun. The asters close at evening, the chicory is entirely shut by noon, the ironweed and golden-rod, un-daunted, spread out their glories through all the hours of the day and night. " Blessings brighten as they take their flight " is a saying that now more than savors of triteness; it holds good of the flowers, which put on their most gorgeous hues with the advancing season. Of these late autumn beauties the chicory is the most useful—it is a roadside bloomer, and many a tired wayfarer has no doubt been cheered by "its eyes of heavenly hue." In France and England the roasted roots are exten-sively used mixed with coffee and ground, the chicory, it is claimed, making the coffee more tonic and nutritious.   Chicory leaves are a favorite salad in France, the gardeners blanching them as we do celery stalks.   Horace frequently mentions chicory as one of the prominent articles of his frugal diet, and Pliny tells us that the Egyptians used it freely on their tables; it is a prominent salad constituent in Egypt.

CHAPTER IX

THE PILGRIMS OF THE YEAR

SEPTEMBER

"The violet loves the sunny bank,
The cowslip loves the lea,
The scarlet creeper loves the elm,
The pine the mountain free."

THE year around, and the world around, journey
the plant pilgrims. Among those perennials which
are found in all latitudes and seasons are the lichens,
fungi, and algæ, or water weeds. In September,
while we wait for the fruits and seeds to finish ripening, let us make some small studies in these three related groups in the vegetable sub-kingdom called the
thallogens.

This sub-kingdom, one of the chief divisions of the
vegetable kingdom, is known also as the class thal-
lophytes. It contains the simplest forms of vegetable
life. Its chief groups are the fungi and algæ, the lichens
being related to both, as if algæ and fungi had united
in one plant, dividing and somewhat changing the
characteristics of each.

At any period of the year you can find lichens in

148

abundance. They cover ragged rocks, dress up old roofs, walls, fence-rails, and dead stumps, especially delighting in the north side of trees. If we examine them through a magnifying glass we shall see that they are made up of cells, laid side and side like little chains of beads, or of cells expanded into short tubes or threads, lying like heaps of tiny fagots. Instead of seeds, lichens have a fine dust called spores, from which they develop.

Lichens are exceedingly long-lived and excessively slow of growth. The lily attains its lovely maturity in a few months; the oaks, elms, pines, become great trees in twenty or thirty years; the humble lichen often lives forty or fifty years before it is old enough to complete its growth by producing spores. Botanists say that the life of a lichen is fitful and strange, and is practically indefinite as to duration. Lichens simply live on and on.

Some lichens have been known to live nearly fifty years without seeming to grow; they appear to dry up, and nearly vanish—then suddenly, from some cause, there is a revival of growth, they expand again. Small and insignificant as these lichens are, they often outlive those longest-lived of trees, the cedar of Lebanon and the California redwood.

The condition of lichen existence is water, for from moisture alone, in dew or rain, they secure their food.

The carbon, oxygen, ammonia, hydrogen, in air and rain, afford them their nourishment. The lichen generally refuses to grow in foul air laden with noxious gases. In the impure air of cities few appear, but they abound in the open country. They absorb by all the surface except the base by which they are fastened to their place of dwelling. They have no roots, and simply adhere to bare rocks, sapless wood, even to naked glass, from which they can receive no nutriment whatever.

In comparison with what is known of plants in general, our knowledge of lichens is yet very limited. They seem to be made chiefly of a kind of gelatin, which exists in lichens only. Humble as they appear, they have always been of large importance in arts and manufactures. They produce exquisite dyes—a rich, costly purple, a valuable scarlet, many shades of brown, and particularly splendid hues of blue and yellow are obtained from these common little growths, which in themselves display chiefly shades of black, gray and green, varied with pink, red and orange cups, balls and edges.

Formerly the gelatin of lichens was used as a substitute for gum-arabic; now other substances have taken its place in commerce. Once it was employed as a basis for perfumes and toilet powders; in this direction also it is out of fashion.

Another use of the lichens is as a food-stuff. The
reindeer of northern lands finds in the lichens under
the snow his chief food. The most nutritious of the
lichens is the Iceland moss, a valuable diet for invalids. This moss, ground to a powder and mixed
with milk, is often the chief nourishment of poor
Icelanders. The Esquimaux, the red Indians, and
Canadian hunters in polar regions use freely the
" tripe lichen " which grows upon rocks in very cold
climates. On the barren steppes of Asia is found
the " manna lichen," growing unattached in beds
from four to six inches thick. It is of a gray color
and sweet taste, offering sustenance to dwellers
where no other vegetable life is found.

Long ago lichens were much used as medicines.
They are harmless, as no poisonous lichen is
known, but they are no longer regarded as remedies.

When the ordinary flower gatherer speaks of liverworts he refers to a member of the crow-foot family,
the hepatica triloba, also called the liver-leaf. This
plant receives its common name from the shape of
its thick, three-lobed leaves, shaded with dark red,
brown, or liver color. The liver shape of the leaf
caused the " simplers," or herb doctors, to prescribe
it in liver complaints. The botanical name hepatica
is from the Greek for liver, also given from the shape
of the leaf. The true liverwort which the profes-

sional botanist recognizes is, however, not a flowering plant, but a lichen, one of the most interesting
of its class. We have all noticed clinging to damp
stones, to decayed logs near a spring, or to the north
side of rocks, something like a dark green, moist, flat
leaf, curled up at the edges. In September these odd
patches are in their glory, and in September, if we
can find a rock from which wells a spring, we shall
be sure of finding liverworts in abundance. Our
first experience will be of the difficulty of removing
them from the rocks to which they adhere, closely
applied by all their under surface except the full,
ruffled edges. As even when a flat, wide knife-blade
is used they are not likely to come off unharmed, let
us sit down by them, lens in hand, and study their
beauties and the wonders of their structure. First
we perceive, when we violently tear one or two specimens from their hold, that the under surface is provided with hairs like spun glass, which lay hold of
the minute interstices and irregularities of the stone
and hold the plants down closer and more firmly
than any cement could do.

Next, turning our attention to the upper surface
of this leaf-like expansion, we find it covered with
dark-green lines, so disposed as to divide the epidermis into a diamond-shaped pattern, which is yet
further beautified by a dot in the centre of each

diamond. Now each of these diamond-shaped spaces is a distinct through and through division of the liverwort expansion, and the dot in the centre represents the throat or mouth of this portion, being the opening by which so much of the plant as one diamond represents breathes out oxygen and other elements of which the space desires to rid itself, and breathes in the carbonic acid gas upon which, in common with other plant life, it feeds. In fact, this dot, which to the unaided eye appears to be merely ornamental, under the magnifying glass reveals itself as a tubular orifice, a ventilating chimney. Having investigated the under surface, with its spun-glass hairs serving as roots for mooring, not for nourishing this plant, and the diversified upper surface with its air holes, let us make a crosswise section, and magnifying it many diameters we shall see that upon the under surface are laid several rows of cells; each diamond-wise division is elevated at the centre, something like a house roof, the dot, composed of minute cells built into a tube, being at the centre. The surface of the liverwort is composed of a layer of small bead-like cells, and between these and the lower cells grow tiny fleshy-branched objects like cactus-plants reduced by a million fold. Dropping the lens we observe the liverworts with our naked eyes, and we see that they have growing upon the

surface small, plant-like objects about half an inch
high, and of a lighter green, grayish, or even of
brighter color. On one liverwort expanse these blossoms, for they are what serve this lichen for blossoms,
are shaped exactly like the little leather-colored
fungi that abound over some spot where wood has
decayed or a hayrick has once stood. Here they
stand, tiny fringe-edged umbrella-like things, and here
on the next liverwort is something quite different!
The growth on this other leaf is also about half an
inch high, and looks as if the fairies had been making
toys and imitating the little tin trumpets which boy-
children love to blow. These atomic trumpets stand
mouths up, but the edge of the mouth is waved or
scalloped. Now bring the microscope to bear again
and these little umbrellas appear to be made of
glittering spun glass, abundant in gay fringes.
These umbrella or mushroom-shaped growths answer
to the pistils of flowering plants, and are nothing
more nor less than seed-discs; that is, spore pro-
ducers. As in flowering plants it takes two to make
a bargain or produce the future plant, the pistillate
liverwort, or what answers to the pistillate form of
higher organizations, must be fertilized by what
represents the staminate growth. That is supplied
by the liverwort which bears the trumpets or clubs,
whichever we choose to call them. We will find

that these do not really start from the upper surface, as do the umbrellas, but from among the glassy hairs of the under surface; then they make a curve among the indentations of the ruffled edge, and pressing upward appear to grow from the edge of the upper surface. Each of these clubs is full of minute pockets, each pocket is full of dust, each atom of dust is like a little whipstock with two lashes; each of these is dependent on water to expand and set it free, and for their convenience the liverwort grows where water abounds. Once visited by water, the lashes of these microscopic whips begin to wriggle and snap, and on some convenient puff of air find their way to one of the waiting umbrellas. Here the cases of spores are fertilized, and when ripe become so active and elastic that they leave their cases and sow themselves around by millions. Of these vast numbers it is but the few which survive, and of those that produce other liverworts the produce may be not in its turn fruitful, but sterile.

One singular fact about liverworts is this, that they may take quite another fashion for reproducing themselves. Some, instead of the clubs and the umbrellas, produce on the surface little elaborately decorated cups of emerald green, sparkling with atoms of that glassy fibre which moors the under side of the leaf to the rock. In the light, and under a magnifying glass, this powdered glass shines like the dust of all manner of jewels, and within the tiny nest lie little spores as eggs lie in a bird's nest. Again water comes to the rescue, dew or rain fills the nests to overflowing, and borne out on the waters go the spores to adhere to some rock surface and expand to other liverworts. Again, instead of such cup-shaped nests full of spores, the upper surface of the liver-wort may produce bags, or flat pockets, or little squat bottles; whatever it is, it is filled with spores, and thus acts as a seed-case, while the naked spores perform the functions of a seed. Such a study of a liverwort, a typical lichen, introduces us to some of the mysteries of lichen life. In it all we are talking about the "infinitely little," but since we have brought our minds to try and conceive of twenty-five or fifty thousand bacteria in line on one inch length of any surface, or hundreds of them easily accommodated side by side on the head of a pin, we can allow ourselves to discourse of liverwort cells, fringes, sacs, pockets, nests, clubs, and whiplashes without flinching. All this is what exists and what we might see if our eyes were strong enough.

FUNGI.

While not so abundant as lichens, the fungi are well known everywhere. We cannot claim, as for lichens, that they are harmless, for many are a virulent poison ; others have a disgusting odor, and nearly all are dangerous in their decay. On the other hand, many of them are a useful, delicious food, and nearly all are beautiful when first developed. Their variety also is very fascinating.

In a walk of less than two miles in a wet summer, may be found twenty different kinds of fungi—some no larger than a pea, some eight inches in diameter. They may be round, oval, flat, cup-shaped, horn-shaped, cushion-shaped, saucer-shaped; they are snow-white, gray, tan, yellow, lavender, orange, dark brown, pink, crimson, purple, and variously mottled ; scaly, or smooth as with varnish. Placed on a large platter among dark green mosses they will be, for one day, a magnificent collection.

One large, egg-shaped variety, growing in pairs, is of a purple shade, very solid, and when broken open seems filled with glittering matter, like iron or steel filings. Another tan-colored, plum-shaped fungus, firm and smooth, is of nearly a royal purple within.

September is a good month for the study of fungi, especially after the early fall rains, when the woods and pastures will be found well filled, not only with brilliant, useless, or poisonous varieties, but with delicious edible kinds. Popularly people call the edible specimens "mushrooms," and the rest "toad-stools," the number of poisonous or of edible instances so named depending rather upon the amount of knowledge of the collector than upon the real qualities of the fungi, for many denominate as "toad-stools" what others know to be an excellent food.

Many varieties not usually eaten are wholesome, and many which human beings reject other animals thrive upon. One large, brown "toad-stool" of the woods is, in its season, the chief food of the wood-tortoise.

The truffle is a rich, expensive food-article, a fungus growing generally under ground. It is much used in flavoring dishes, and is never plentiful, as demand always outruns supply. Dogs and pigs are trained to hunt truffles by the scent. The animals run along with their noses to the ground until they find a truffle bed. They are rewarded by a piece of cheese, which encourages them to continue the hunt and not dig up the dainty for themselves. The truffle is roundish, dark colored, rough skinned, white in the interior, and of strong odor. Some truffles are strawberry or apple scented, others have a decided smell of onion or garlic.

In general a fungus may be defined as a thallophyte without any chlorophyl or leaf-green in its composition. Among the brilliant colors displayed by fungi no green or blue can be found.

The most popular and useful fungus is the table mushroom. This rarely ever grows in the woods, in shade, on wet lands, or on decaying stumps. It prefers the open, breezy, well-sunned pastures, where the grass is kept short by the grazing of sheep or cattle. Early in the morning, or shortly before sunset, the dainty white or cream-colored buttons, borne on snow-white stalks, push up through the soil, and gradually expand until the discs are flat or but slightly convex. From two to six inches is the diameter, seldom more than two or three.

Varieties of the pasture mushroom are few, and can readily be learned. The mushroom is composed of stem and cap; the stem is finger-shaped with the roundish end in the earth. About half way up is usually a ring of the covering skin, where, in the button-shape, the veil of the mushroom was attached.

This veil extends over the cap, and is left at the edge in a little frill; it can be easily stripped off. Under the veil the flesh is ivory white, and is smooth and firm.

The under side of the cap is laid in plaits called gills, from their resemblance to fish gills. They never grow fast to, or down upon, the stem, usually stopping short off, about one-tenth of an inch from its juncture with the cap.   Mushrooms are cultivated in gardens or cellars.   They grow from spores and little finger-like lengths, called spawn, which are produced by the spores.   Mushrooms turn black or purplish after the first twenty hours of growth.   When the gills have taken this dark hue the mushroom is unfit to eat.

Some fungi grow in very wet places; the woods are likely to be full of them after a few rainy days. They are all short-lived.

One or two fungi have phosphorus in them, and shine in the dark.   Fungi are generally parasitic on some underground decay, as they lack leaf-green to assimilate food for themselves.

While lichens seem to promote the growth of plants upon which they are fastened, fungi generally promote decay and hasten death.   The lichen seems to be a link between this class fungi and the algæ, or water weeds.

William Hamilton Gibson is probably the dis-coverer in this country of a very wonderful fungus, which thus far has only been described by himself in articles in the " Scientific American " and " Harper's Young People."   The marvel about this fungus is its carnivorous habit of growth.   From his account of the specimen observed we draw our statement.   Near the bank of a stream, in a bed of rich green moss,

Mr. Gibson saw a brilliant orange-colored cone, not quite an inch in height. Examination showed this to be the cap of a small, inedible mushroom. The little cone had a long stem penetrating deeply the moss bed, and Mr. Gibson, knowing that all fungi grow upon some buried body—vegetable as a matter of custom—concluded to dig the orange-tinted fungus up, and see what manner of food it was using. Separating the moss and working with care he secured the entire fungus, and found it fastened upon a chrysalis an inch and a half in length! The roots of the fungus branched and penetrated the chrysalis in several portions, being there firmly imbedded.

Having dissected the chrysalis, Mr. Gibson found in it the perfectly formed moth, which had reached maturity and been ready to emerge when the roots of the fungus penetrated its body, sucked up its juices and accomplished its death. The moth had been ready to seek air and light on a pair of wings, a handsome insect; instead it was now reaching air and light in the form of an orange-colored mushroom, whose period of life would probably be limited to no more than forty-eight hours, if so many.

Mr. Gibson assumes that this was no erratic fungus, living in a manner singular and individual, but was of a family of feeders upon animal life, and if

11

an instance were found again it would be rooted in a chrysalis. He states that it very plainly belongs to a genus of fungi, of which several species are known and described, all having this habit of transforming buried insect life into fungus tissues. In New Zealand one of these fungi plants itself upon the head of a caterpillar; always upon a caterpillar, and always upon the head. It rapidly attains a growth of several inches, absorbing in its increase the body of the insect. Nutriment from this source failing, the fungus extends its roots to the soil, and there completes its growth and decay. In the shops of apothecaries in China a prominent article, considered a specific for many ills, is sold in little bundles. These bundles appear like a bunch of dry and cooked sticks four inches long. Examination shows that each stick is a dried sample of this fungus which thrives on caterpillars, and at its root end is attached the equally dried head of the unfortunate insect.

The singular circumstances of this growth, this strange union of an animal and vegetable, no doubt impressed the Chinese mind as a manifestation of some peculiar power, and convinced them that necessarily it must possess healing virtues. Nor is such a notion without parallel in the English race. There is an ancient medical work which has a " cure

for idiocy" in an oil called "oil of manskull."
Thyme, rue, lard, and the ground skull of a recently
dead person were to be mixed and melted into an
ointment, which, rubbed into the idiot's head, would
produce excellent results! The Chinese with his
fungus-caterpillar was no more absurd.

ALGÆ.

Let us look at the algæ. Unlike the fungi they
have a portion of chlorophyl, or leaf-green. They
are known as fresh-water algæ and salt-water algæ
(usually called sea-weeds). This suggests to us that
their homes being in the water, they must secure
their food, as do the fishes, from the water and the air
contained in water. This accounts for their being
in most cases so very finely divided. Nearly all
leaves or fronds growing under water are much
parted, as if they had been cut into little strips or
fringes.

The algæ are flowerless plants, yet from their delicate, graceful forms, and lovely coloring, are often as
beautiful as flowers. The gardens of the sea, with
their flowerless plants and flower-like animals, the
sea-anemones, are as charming as the gardens of the
land.

The fresh-water algæ are mostly green, pulpy, and
not so richly varied as sea-weeds. Some of them

grow in hot springs, where the water is so warm you can scarcely bear your hand in it; some of them, the tiniest of all, the red-snows of the arctic regions, grow upon snow fields.

The sea-weeds offer us most pleasing studies. The broad leaf-like, sometimes much-branched part is called the frond, also the thallus. This may have a stalk and a root-like expansion; these do not serve any ordinary purpose of stem or root, they merely fasten the weed upon some stone or shell, and permit it to wave gently about in the water. In this waving motion the algæ are fishing, as the barnacles do, gathering food from the water and the air in the water; for, like all other plants, their food is mineral substances held in solution, and gases gathered from the air, such as carbon dioxide and oxygen.

Some sea-weeds are merely straight, coarse grass-like blades, as the abundant California weed; others are coarse, of a dull olive green, much branched, like the bladder-wrack which often lines the shores. Others are broad, curly fronds of green, red, brown, or nearly white, as the various kelps. The waters of the ocean are sometimes so full of a moderately coarse red weed, like a tangle of crimson strings, that the great waves lose their green color and white, foamy curves, and break upon the beach like waves of blood.

Bread Mould

Sea-weed.

Cheese Mould.

Mushrooms

Lichens

Nina G. Barlow
'98

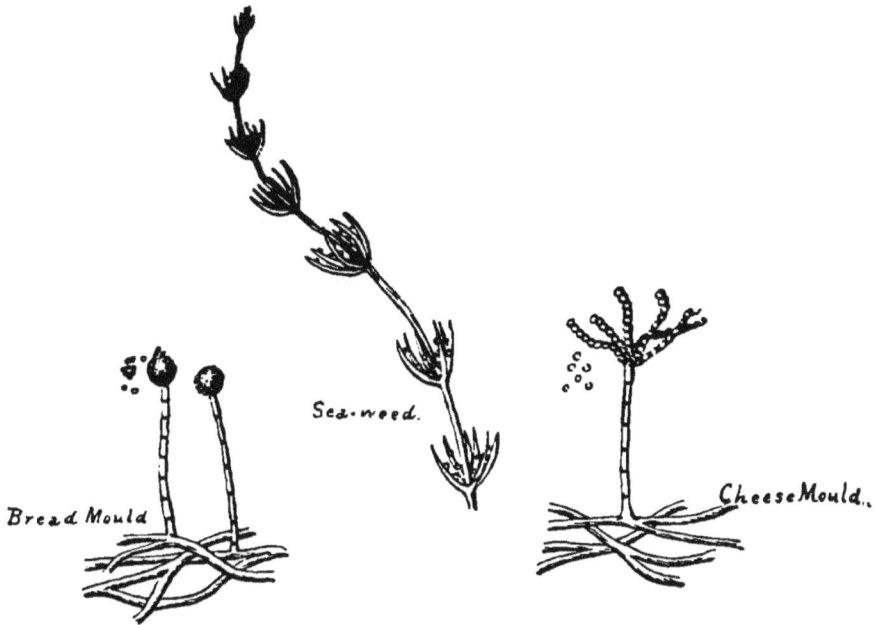

THE FLOWERLESS PLANTS

Many sea-weeds are very finely divided, graceful and exquisitely formed, beyond all other plants, and these lace-like fronds, beside which maiden-hair fern and the fringe-tree look coarse, are tinted in brown, red, pink, crimson, green, and minglings of these colors.   Blue, purple, and pure yellow are shades unknown to sea-weeds.   The nearest approach to yellow is the bronze of the tiny ball-bladders of the gulf-weed.

Algæ are made of cells held together by a kind of gelatin.   Sometimes the cells elongate so as to be tube-like.   In some cases the cell-walls are so clearly marked that they look like strings of beads.   Like other flowerless plants, sea-weeds grow not from seeds, but from spores.

Corallines are stiff sea-weeds, coated thickly with lime gathered from the water.   Some sea-weeds, as dulse and Irish-moss, are eatable.   The largest plant in the world is a great sea-weed which grows to be from three hundred to seven hundred yards long.   Gulf-weed, or sargassum, has never been found fastened, but floats in masses, often of several acres extent.   The Sargassum Sea is a mass of this weed many miles broad spread out upon the water. The stalks of some of these great weeds are larger than a man's body.

Many fishes and other sea animals feed upon sea-weeds. Algæ are of some value as fertilizers when spread upon the fields; certain kinds of kelp were once the chief source of our soda supply.

PARASITIC PLANTS.

These are plants which do not draw their chief food supply directly from the earth or air, but lazily settle upon other plants, and feed upon what their more industrious neighbors have converted into food stuff. Plant idlers and paupers these.

The mistletoe is parasitic, as it grows from a seed which being dropped upon a branch takes root there, and derives nutriment from the sap which it sucks by its roots, forced into the fibres of the tree-host. The mistletoe is not a true parasite, because it retains its green leaves and by these gathers from the air and digests a large part of its food. The mistletoe matures a pure white wax-like berry, about as large as a chick-pea, or big barley grain. It is an evergreen plant, and our Saxon and Celtic ancestors considered it sacred to the gods and endued with mystic power.

A true parasite is the dodder, a slim vine which climbs about small plants. It has clusters of little pink flowers, which look exactly like tiny morning-glories, or wild bind-weed blossoms. The stem is like a long pink thread, and here and there puts out rootlets, which work into the plant to which it clings, and drink its food stuff. The pink blossoms attract insects to the dodder's bells to carry pollen and fertilize its seeds.

Long ago, no doubt, the dodder had green stem and leaves, and climbed, as does a morning-glory; but as it by degrees took to living upon the work of other plants and ceased absorbing and digesting food for itself, the leaves ceased to develop and the chlorophyl no longer formed.

We might take the dodder for the text of a little sermon about idle people who will not work and insist upon being supported by other and better folk, never caring if these are worn to death by the burden.

The Indian pipes or beech-drops are all parasitic; their chlorophyl is gone, and their once green leaves have turned to little bleached scales. There is good Scripture for their being served that way. The one unused talent must be taken away and given to him who used five or ten talents well! These pale idlers under the pine trees and beeches were once good, busy members of the wintergreen family.

Broom-rape or "broom thief" is a parasitic plant, with a tall strong stem, scales, and tiny bloom, all a dull red-brown; a "seedy"-looking plant-idler that has taken to loafing round for its plant neighbors to maintain.

Of late much attention has been paid to the study of parasitic plants. The English call the invaded plant the host, and speak of a parasite with some respect in spite of its disreputable way of obtaining a living. The mistletoe has, in a manner, conferred dignity on parasites in the English idea, because the mistletoe was held sacred by the Druids. Naturally this mistletoe does not choose the oak, but the apple; the Druids transplanted it to the oaks because the oaks were objects of worship and were tree temples. Mistletoes, as we have seen, are only partial parasites, and other instances of partial parasitism have but lately come to notice in what are now called clandestine parasites; these are plants whose tricky habit was hidden, and for long was unsuspected. Among these, the error of whose ways has been but recently discovered, are the toad-flax, wood betony, eyebright, and cow-wheat. Gerardias are parasitic upon roots of other plants, but have a well-developed leaf-surface gathering food from the air. One of the largest flowers in the world, sharing the palm of size with the great Victoria Regia lily, is a parasite found in the Island of Sumatra—the monstrous Rafflesia Arnoldi; this is a direct parasite, has nothing like leaves, is merely a flower growing directly on the stem of a Cissus—a flower only, but what a flower! It weighs nine pounds, and is of the circumference of a cart-wheel; its very appearance betrays its villainy. Its color is livid, its odor is that of carrion, and it steals the life of its unhappy host, offering neither beauty nor fragrance for the hospitality wrested from it.

Saprophytes are parasites which live only upon decayed vegetable matter. Their host is not a living, but a decayed tree. Of these, the beech-drops, pine-sap, and cancer-root are examples.

CHAPTER X

OCTOBER

"It was autumn, and incessant
Piped the quail from stocks and leaves,
And, like living coals, the apples
Burned among the withering leaves."

OCTOBER is the chief fruit month of the year. In the woods the walnuts, chestnuts, and acorns rattle to the ground in the frosty mornings. Along the roadside, hedges, and old fences the bitter-sweet opens its stiff red and orange cups, while hazel-nuts drop from their protecting cases. The corn stands in shocks up and down the fields, and heaps of the newly-husked ears glitter golden in the sun.

The leaves have withered on squash and pumpkin vines, but the great bright globes of their fruit lie still ungathered. The last watermelons and citrons, their deep green streaked and mottled with white, are being carried to market. There are more ruddy apples than leaves on the trees, while the fruit stores are full of peaches, plums, grapes, and pears.

170

The machines are busy threshing out wheat, oats, rye, barley, clover, and buckwheat. Every weed and wayside bush is covered with ripe seed. Truly the earth has brought forth her fruits.

What is fruit? Ask the dictionary, and the reply is, "The matured seed vessel, its contents, and such parts as are incorporated with them." Ask the child; he answers, "Something good to eat uncooked."

Fruit is the matured seed; all nuts, grains, peas, beans, the winged seeds of maple, ash, thistle, dandelion, are fruit. Fruit-bearing is the life-object of the plant; the fruit contains the embryo of the plant to come, and assures the continuance of vegetation and of animal life.

More narrowly, fruit is defined, as we generally understand it, as "the pulpy mass connected with seeds of various plants and trees, especially such products as are pleasant to the taste and are eaten by men and animals." This is the popular idea of fruit, and we recognize oranges, lemons, pine-apples, peaches, cherries, and their kind as undoubted fruits. It requires some consideration to enable us to accept as fruit the slowly-floating dandelion seed with its little silken sail, or the sharp-hooked burrs of the stick-tight, or burdock, or stramonium.

The word fruit is from the Latin *fruor*, "I enjoy." We cheerfully accord it to the beautiful and luscious, denying it to the harmful and unsightly. Nature differs with us here, molding and maturing with the earnest care the fruits of her unbeautiful children, which we style weeds. Let us courageously follow both science and nature, looking at fruit in its widest sense.

Fruits are divided into three classes : I. Fleshy ; as berries, pears, gourds, melons, grapes, apples. II. Stony (drupaceous) ; having a stone or pit, as peaches, plums, cherries. III. Dry ; as nuts, peas, beans, grains.

Among the dry fruits we find that abundant nut, so singular in its manner of growth, joy of the American boy's heart—the peanut. This low-growing vine-like plant bears small yellow blossoms, shaped like bean flowers. The calyx is very long, and the germ with the ovules lies at its bottom. The style of the pistil is also long. When the flower has dried and dropped away we see the ovary expanded to a pod with from one to three seeds in it; this is borne on the tip of a flower-stalk, which at once begins to grow rapidly, bending toward the earth. This stalk bears the several pods of the flower cluster, and as if pressed to the earth by their weight, touches the ground and begins to push into it, burying the pods. When some few inches under the soil the pods harden, becoming woody, and the seeds mature.

Each seed is composed of two thick cotyledons and a straight germ, lying in the fibrous shell. These oily, richly-flavored shell-protected cotyledons are the peanut of commerce.

The size of seeds and seed-pods often attracts even the casual observer. How large are the pods which spring from the little blossom of the lima bean! The small flower of the honey-locust expands from its tiny germ, growing the summer long into a great curled pod over an inch wide, and from seven to ten inches in length. How small is the dainty apple blossom! That little green knob below the five partings of the calyx grows to the fragrant, juicy, richly-tinted apple often four or five inches in diameter. The cocoanut and the Brazil-nut are seeds of a notable size; contrast with these the tiny seeds of poppies and the portulacca seeds, which appear as a fine steel-colored dust.

Many seeds which are small in themselves have very large cases. The pumpkin seed is a thin oval, at most three-fourths of an inch long and half an inch broad. From it sprouts a vine which may be some yards in length, bearing several large pumpkins of from three to twenty-five or thirty pounds in weight. Some pumpkins are so large that they may be cut into chariots or slippers, in which a five-year-old child could be placed to play Cinderella's god-mother, or the " Old Woman Who Lived in a Shoe."

A very remarkable compound fruit is the pine-apple. The flowers of the pine-apple are small, growing in a dense spike, guarded by prickly leaves, upon a short, thick stem. When the faded flowers drop off, the ovaries enlarge greatly, becoming soft and juicy, and growing together into one solid fruit, cased in the thick scales of the united ovary cases. The pine-apple seeds are inconspicuous, buried in the fragrant flesh of the ovaries. Thus one large fruit with many tiny seeds is produced from a number of very small flowers.

A pine-cone is a fruit similar to a pine-apple in its growth, and a young green pine-cone looks much like a young green pine-apple, except for its crown of green leaves. Both the pine-cone and the mulberry resemble the pine-apple in being multiple fruits— that is, many ovaries united to form one fruit. The pine-apple ripens into juicy richness, the pine-cone remains dry and woody, its hard-closed scales finally loosen and bend back, allowing the small seeds to drop out.

Another curiously formed fruit is the fig. This is so common an article of commerce and food that we generally eat it with very little thought of the marvel of its growth. The fig stem expands at its axis or summit, and as its walls swell out the interior is left hollow. In this space a cluster of flowers develops and matures seeds, the flowers becoming a thick sweet pulp, in which the numerous very small seeds are contained. The expanded stem serves as a rind for the fruit thus formed, and itself becomes rich, sweet, and tender, the whole composing the luscious fruit.

In strawberries, blackberries, and others, the minute seeds are held together by pulp forming a compound fruit. The minute yellow spots on the crimson surface of the strawberry are the seeds borne on the modified receptacle, which, for purposes of seed distribution, has become highly colored, fragrant, and richly flavored.

Seeds are of almost infinite variety in shape and color. Any seedsman's store will exhibit to us red, purple, yellow, white, black, gray, brown, pink, green, and gayly-mottled seeds. As for shapes they are round, oval, flat, discs, stars, horns, crescents, shuttle-shaped, spindle-shaped, full of Protean surprises in their change of form. All these shapes, colors, and flavors have a meaning in the matter of seed distribution. The seed having come to be, must secure its planting and germinating.

Some seeds have pods which, when fully ripe, spring open with such sudden violence as to cast the seeds for several yards about the parent plant. The touch-me-not of the gardens is a good example of this, also the jewel-weed of the brooksides.

Other seeds are provided with wings, as those of the maple, which grow in pairs so that there are two wings, or as the ash, where each seed, or key, has one wing or sail, so it can be borne as a shuttle-cock.

Thistle, dandelion, milk-weed seeds are furnished with silken sails by which they are carried for long distances by even the lightest breezes.

Seeds with hooks and burrs depend for transportation upon animals, to which they cling, and having been carried about for some time, they are finally dropped where they can germinate, far from the parent plant. Sheep, cows, and horses in pasture may often be seen with hundreds of " weed-seeds " cleaving to their hairy coats.

Certain seeds have stiff projections, sensitive to moisture and heat, by means of which, as by screws, they work their way into the earth. This singular power was in old times deemed a species of witch-craft.

The number of seeds is as countless as the sands on the shore, the stars in the sky, or the drops in the sea. How many seeds mature in a single head of clover or panicle of grass! How very many are the seeds of a single melon or pumpkin, or poppy head!

Who thinks to count the seeds in a single head of dandelion or thistle? What millions of seeds drop from pines, elms, oaks, maples!

The spores of the flowerless plants float off in clouds, millions of minute but life-full atoms. How seedy we consider lemons and oranges; how very many are the seeds held in the abundant crops of berries and currants!

Were seeds not infinite in number the plant-world must perish, seeds being exposed to so many disasters. They are destroyed by over-much water, decaying as they lie in or upon the ground. Exposed, as they frequently are, the tender germ is killed by freezing; hot suns burn and dry the life out of millions more. They are devoured by beasts and insects; in many seasons almost the entire walnut crop is destroyed by worms, come from eggs deposited by an insect mother in the young and tender nuts in their early green state. For several consecutive years insects will destroy the fruit crop.

When the seed has finally matured, been preserved from such mischances, sprouted, sprung up, further accidents await it. Men and beasts may crush it by treading upon it; beasts eagerly graze upon the new plantlets; storms and suns may prove as disastrous to the fresh growth as to the seed. When so many seeds are doomed to perish, a vast

12

number must be produced in order that enough shall survive to be mature seed and develop plants.

The plant then secures its continuance by means of vast numbers of seeds. Advancing further in the scheme of vegetable preservation, in order to secure favorable localities for growth, the plant forms partnerships for seed distribution, as it did for the fertilization of the ovules.

Who are the chosen partners? Not insects. Insects are enemies of the seed, although allies of the flowers. The first partner chosen is the wind. The wind carries abroad all seeds provided with wings or sails, and small seeds light enough to be swept out of their scales, or open cases, by a brisk breeze, and be by it scattered abroad.

Another partner is the waters. What thousands of cocoanuts and other tropical fruits, provided with waterproof cases, have been carried by the waves to clothe with verdure waiting islands. How are the tribes of lilies and rushes and arrow-plant, and sedge, and a hundred others, distributed by the waters of streams. From island to island, from continent to continent, from stream to stream, lake and pond, go the water-carried fruits of the earth, with their blessings and their beauty.

Animals are inveigled into partnerships by plants. Many seeds, notably those of red sorrel, plantain, grasses, are conveyed from place to place by grazing beasts. The seed is contained in so hard a husk or case that it is not digested with the soft parts of the eaten plant, and is finally dropped in conditions promotive of rapid growth.

Squirrels, field-mice, and other little rodents that have a habit of storing up winter food are fine seed planters. It has been estimated that, if let alone, squirrels would each year replant many acres of woodland, especially with such valuable trees as the cherry, oak, chestnut, walnut, beech. These squirrels carry off and hide or accidentally drop thousands of nuts of which they forget the hiding-places.

Of all the plants' partners the birds are the most helpful. For them the fruits put on their richest colors, and take their juicy sweetness, their perfume. Seeds, berries, all small fruits, as currants, gooseberries, cherries, bid for the aid of birds in seed scattering. The bird often swallows the whole fruit, as the thorn-apples and rose-haws. The pulp and flesh are quickly digested and the hungry bird is ever seeking for more. The seeds and pits are too thickly coated to digest, and are dropped as the birds flit from place to place.

We notice the hardness of the pits of peaches, plums, cherries; the firm, horny cases of lemon, orange, apple, and pear seeds, assuring them against rapid decay. Of such fruits the bird partner desires only the soft, sweet, pulpy covering, and this he gets as his share in the seed-planting venture.

Most of our choicest fruits and nuts owe their especial richness and size to the careful cultivation which produces valuable qualities from common, hard, acrid, wild varieties.

Although October is the "month of fruits," there still linger flowers to grace the waning year. Not only are bold and hardy plants left to defy the first cold, rough autumn winds—one of the most delicate and beautiful of our wild flowers crowns October days, lingering after its natal month of September, as if loath to leave the world to gloom. The poets love the gentian—some call it "blue," some "purple;" whether because they are color-blind, or because the word chosen better fits their lines, who can tell? Thus Bryant says:

"Blue, blue, as if the sky let fall
A flower from its cerulean wall."

And another says:

"There came a purple creature
Which ravished all the hill."

This lovely blossom, the fringed gentian, grows on a stem from one to two feet high, branching at the top, each branch bearing a funnel-formed corolla of four fringed lobes. The calyx is also four cleft, and there are four stamens; the pistil has two stigmas and one style, which is therefore a single style formed by the growing together of two, the stigmas not uniting. Now as the gentian blooms so late its seeds are very late ripening, and, falling into the ground, do not assent to a sleep so short as that they must waken early the next spring. On the contrary, as is the case with hollyhocks and some other plants, the seeds will not produce blossoms until the second year, and the plant is biennial. One peculiarity about the gentian is that it changes its haunts, seldom growing two or three generations of plants in the same locality. The hard, round seeds are wanderers; they are easily washed away, and especially if they have ripened on a hillside, are likely to find their lodgment in a valley or meadow.

All the gentians bloom late. The closed gentian is another blue flower which unfolds in September and lingers to grace October; it receives its name, " closed gentian," from the shape of the corolla, which looks as if always in bud; it is a large oblong corolla of a deep rich blue, the mouth firmly shut, seeming to guard the inclosed stamens and pistil from the dampness and the chilly air of autumn nights. The leaves of the closed gentian are much larger and coarser than the delicate opposite leaves of the fringed variety; the flowers often develop in the axils of the leaves, seeming very cosy and well sheltered by the large clasping leaf bases; flowers also develop in a close cluster at the summit of the stem, and there a number of leaves crowd about them. Both the closed and fringed gentians become of a much lighter color when growing in deep shade.

All along our eastern sea-coast the five-flowered gentian graces the autumnal days. This has smaller and paler flowers, and these appear to be always on the defensive, as every lobe of the corolla is armed with a sharp bristle.

Two very delicate flowers appear in October, having sent out individuals to bloom in August and September. The grass of Parnassus, tall and slender, with little round clasping leaves; this is no grass at all, but so named from the lightness and grace of its stem. The small white or cream-colored flowers, with deep blue veins, look like memories of spring-time.

Also, here is a little member of the great orchid family, a dainty flower which looks like a work of silver filigree—the ladies-tresses. Some botanists claim that this is an inhabitant of low, swampy lands; others assign the hillsides as its chosen home. The

fact is, this plant seems to be well suited with almost
any place or month for blooming. From July to November it may be found in dry woods and pastures,
by dusty roadsides, along the hills, and down in the
swamp edges. The stem is very remarkable; it
looks as if two stems had been closely twisted together, like a little green rope of two strands. The
leaves are long, linear lance-shape; the lowest are
the longest, and all are sheathing; the flowers are
small, silvery white, closely crowded in the slender
spike, and have the delicate fragrance of new-mown
hay. In New England this little orchid is sometimes called the "wild hyacinth," but though it is
wild enough, refusing to abide in gardens, there is
nothing of the hyacinth about it. This is the last
orchid of the year, and also the smallest of our native
orchids. The procession of the flowers almost began
with our largest native orchid, the purple lady's-
slipper.

The bees seem to know that their time for honey-
gathering is now very short, and they throng about
these silvery ringent corollas. When the noonday
sun shines warm, the soft, lazy air is full of the perfume of the ladies-tresses, and everywhere echoes
the hum of bees, we might be beguiled into the fancy
that the year had, like the dial of Ahaz, turned again
backward, and that instead of bearing down on snow

and ice we were re-borne on some tide of good fortune toward midsummer!

Some of the violets linger late or take a fresh season of blooming here in October; the cinquefoil also unfolds its twinkling stars, and the brave yarrow holds itself up in sturdy fashion as if defying fate. The asters have departed, except some late starveling specimens, but the dauntless hawks-weed spreads out just as brilliant gold as in the early spring. The hawks-weed is a well-loved companion of a year, and it and spring go journeying around the world together!

Near the coast the pink knot-weed smiles out, and we scarcely wonder that good, loving, poetic Thoreau thought them bright as a " peach orchard in full bloom."

A flower not unlike in appearance to the closed gentian lingers into October days—the great blue lobelia. The varieties of the lobelia are strangely unlike each other. In early summer we have the *lobelia gracilis*, a fairy of a flower, delicately blue with a white bar on its lower lip. The magnificent cardinal flower of wet lands, superbly beautiful, one of the most splendidly colored of all our wild flowers, seems too haughty and stately to claim kin with the shy, modest *gracilis*, or this strong, sturdy, farmer's boy of a great blue lobelia. This flower has a re-

markable arrangement of stamens and pistils; the
pistil has a fringed stigma, and is fully inclosed
by the stamens, as if they would force their pollen
upon it. We find, however, that the sensitive or
sticky surfaces of the stigma are pressed together so
that no pollen can reach them, while the anthers on
their part open only through a pore at the tips.
This pore does not open until jarred by a visiting
bee, when a sudden snap casts the pollen over his
coat; the anthers being thus relieved of their burden
lift themselves and the pistil seizes the opportunity to
pass beyond them and be on the lookout for the
next honey-seeker who comes along bearing pollen
on his raiment.

The fourth member of our lobelia family does not
linger to greet the fruit-scented October air—the
inflated or puffed lobelia dropped its small purple
flowers long ago. The Indians used its coarse leaves
for tobacco. The great blue lobelia has a sister-
plant, exactly like itself, but with blossoms of pearly
white; there are some other varieties which prefer
moist woods or brooksides.

One of our most magnificent, composite flowers
flourishes in October—the "blazing star." This
flower clings to the low hills near the coast; its
stems are tall, furrowed, narrow-leaved, and set for
the upper half of their length with large, red-purple

flowers. The buds are a deep, fiery purple, and the plant is in its beauty when some of the lower blossoms are expanded, and the whole portion above is set with these rich, shining buds. Unfortunately the expansion of bloom proceeds slowly toward the apex of the stem, and leaves behind it withered brown fringes that were "blazing stars," and then rough seeds, fruit for these October days.

What is this we see in the hedge as we go homeward? A rose, a wild, pink rose! Every now and again these late-blooming roses, sweet and bright as in June, surprise us in our autumn walks.

CHAPTER XI

NOVEMBER

" They know the time to go !
The fairy clocks strike their inaudible hour,
In field and woodland, and each punctual flower
Bows at the signal an obedient head
And hastes to bed."

EVEN the youngest and least educated observers of flowers understand that they have fixed seasons of bloom and decay, of folding and opening. The child soon learns to look for chick-weed, dandelions, and butter-cups early in the spring, to expect roses and lilies in June, hollyhocks, asters, and petunias later; to associate dahlias, chrysanthemums, and golden-rod with autumn.

Bands of school children seek in April the pastures and woodlands for wild hyacinths, violets, anemones, blood-root, and spring beauty. The least astute gardener prepares his crocus, hyacinth, tulip, and pansy beds for earliest bloom. The flowering almond, peonies, iris, columbine, · garden valerian, come promptly in the train of May ; the yellow rose is the first of the June band to unfold its petals.

187

With the return of mild weather, sun and moisture set at work the perennials; the root-caps under ground send up stems and leaves. Most self-sown seeds germinate the spring after they ripen; some require two seasons to develop the embryo; others will sprout, sending up plantlets within a few days or weeks after they are buried in the soil. One plant is much slower than another in the growth of stem and leaf, more tardy in unfolding blossoms; all this assures the succession of the plants, the slow march past us of the beautiful hosts. When plants are crowded or hurried out of their natural time of bloom they seldom give as strong, lasting or fine flowers as when they are allowed to take their natural course. Flowers which constantly blossom in winter must rest in summer. Even those known as "perpetual bloomers" are by no means always in flower. Geraniums for winter flowering must be trimmed and shaded in summer to keep them from budding; the oxalis and lilies that are destined for house plants, must have their profound rest in summer, being kept nearly dry and not permitted even to form leaves.

In November the early bulbs are set in well-prepared beds and covered with straw or leaves. The gardener bestows upon his charges the attention that nature herself accords to her wild children. The hardy plants of forest, pond, and field remind one of strong children who bravely take care of themselves while others of their age are still in their nurse's arms.

The water lilies having finished their bloom draw down under the water, giving no sign of their existence save a few faded dull-red pads idly flapping upon the surface of the pond as autumn winds pass by. The wood flowers, when their seed is matured, disappear entirely, not a stem or dead leaf is left to hint where star-flower, spring beauty, wild pansy, white-hearts, or Solomon's seal has been, or will again be. November counts almost no flowers under her cold gray-blue skies and upon her bare hills, unless, perhaps, a few hardy compositæ linger, such as a wild aster or weazened spray of golden-rod, or corymb of yarrow, or a despised little May-weed, making mock of its own name, gallantly flowering vigorously every week from May until snow flies.

As judicious mothers summon in their little ones, and tuck them safely into bed before night has fully fallen, so the loving mother nature folds away her flower children before the night of winter has well begun. She covers their sleeping forms with such blankets of leaves as the robins brought to the "Babes in the Wood." We see no more of them abroad save a few disobedient stragglers, who already look worn, hungry, and aweary from not following the good custom of the family.

In this chilly month of November the roots are resting and sap has nearly ceased rising in the stems. Now is the time to trim vines, trees, and shrubs, when there shall be no plant strength wasted in leakage. Nature here sets us an example; her sharp winds whistle through the forests and snap off twigs, branches, dead pods, sometimes remorselessly taking away great limbs. Pruned, bare, quiet, the vegetable world looks as if it were dead. It is merely asleep. Nature is the grand patroness of rest.

"Blessed," said Sancho Panza, "be the man who first invented sleep!" If ever there were such a man, he was merely following in the wake of good mother nature, who not only decrees a daily sleep for all living things, but in the matter of her plant children a long seasonal rest as well. Even in the tropics the vegetable world arrests its activities, has its quiet time, although the name of that time is not winter.

What is that daily rest which nature gives to the plant? The sleep of plants is a study full of interest. Among the plant motions, one is that of getting ready for slumber. Touch a sensitive plant. Wherever you touch it the plant seems to shrink, the leaflets fold up in pairs, and the stem of the leaf bends earthward. The leaf suddenly appears faded. Watch this same sensitive plant at sunset; you will see it making the same motions, drooping and folding, preparing thus for night, and so continuing until morning.

Walk out on a spring evening. Where are the dandelions which, in the morning, made the grass like a " Field of the Cloth of Gold?" Search for them. They are hard to find on the green turf. The sepals of the wrapper have folded over the closed golden discs. What! Every one faded so soon? No; only asleep, to open in renewed beauty when the day breaks and the shadows flee away.

A field of clover presents a similar spectacle; the leaflets hang back against the stalk as if dried by a hot gale. The sorrel and bean also drop back their leaves and assume the most wretchedly wilted appearance. Never mind, the morning sun will call them back to vigorous life. Folding together, falling backward, lifting into a cluster, the leaves have a variety of ways of getting ready for sleep, just as some people sleep with their arms over or under their heads, laid by their sides, or folded over their chests.

We are all accustomed to seeing the morning-glories close between nine and ten in the morning, while the four-o'clock does not wake from sleep until late in the afternoon. The beauty of the night unfolds near sunset the cream-colored petals that guard its purple heart; the water-lilies unclose at night, and shut their silver chalices by high noon. Children have nicknamed the chicory " Miss Go-to-bed-at-noon." The evening primrose never opens until the sun is low; it sleeps by day and wakens by night, loved of the moonlight, the whip-poorwill, the nightingale, and the moth. Tulips and marigolds will be found sound asleep by dark and do not waken until the sun is high.

The sleep of flowers rests the organs of growth; it serves also in the folding of leaf and blossom to protect the plant from chill and to shelter the stamens and pistils from the heavy dew. For some plants the day-closing shuts out insects which might injure them, and the night-waking calls night-flying insects to distribute the pollen.

Cloudiness or coming rain also cause many flowers to close or to remain closed. Tulips remain fast shut on rainy mornings; were the large petals wide open the flowers would be ruined. Marigolds, morn-ing-glories and many other flowers refuse to unfold unless the weather is fair. Plants are thus both clocks and barometers.

The trunk of the tree bears in its rings the story of the years through which it has lived. The stalk

GOOD NIGHT

of the sunflower writes in its twisted fibres the tale
of sun-risings and settings which it has followed;
the flowers and leaves are chroniclers of the hours of
the day and the hourly changes of the weather.

The small scarlet pimpernel on the roadside attracts little attention; put it under a microscope it
becomes a thing of marvelous beauty; its small
gray-green leaves seem to be powdered with frost;
its scarlet petals are delicately veined; the dark
centre glows into a round of royal purple. This exquisite flower is one of the best of weather prophets.
Hours away it knows of the coming of the rain,
folds tight its petals so that scarcely a red tip shows,
bends down its leaves, and thus exposing the least
possible surface, waits for the approaching storm.
The later any of these sleep-taking plants linger in
the autumn the more wakeful they become. They
are like children—made drowsy by the hot summer
day, but alert and wide-eyed when the days are crisp
and cool.

Perhaps there is no flower more fascinating in its
opening than the splendid moon-flower, a radiant
creature of Flora's own, which might be described
as the Sultan's daughter in the "Arabian Nights"
—"round-faced and beautiful as the full moon in the
seventh night." The moon-flower, from a bulb,
sends up a most luxuriant vine, with leaves resem-

13

bling those of the morning-glory, but larger, more glossy, and finer in texture. These clothe with a wall of verdure any trellis which may be set for the accommodation of the plant. All the summer day this deep rich green lies under the sun, as if rejoicing in its rays, and parting the leaves gently we may find great numbers of buds, each one a calyx holding a long tube, which expands or inflates at the top, with the fullness of the infolded salver-form margin white, a white tinged as if the shadow of a green leaf transfused with light had fallen upon it. At five o'clock, visit the moon-vine; the buds are not now hidden; they extend beyond the leaf shelter; the tube has elongated; the puff of dazzling white silk at the top is larger and clearer. Now watch them well. This is as if one stood by a cradle and watched the waking of a babe. The breast heaves with deep breaths; the muscles quiver; there is a slow rhythmic palpitation through all the resting form. So it is with the moon-flower. The bud trembles; it seems now no mere plant, but waking animal life. There is a quiver through all the folded involucre; the long tube breathes, we see it expanding as in a sigh. Now that folded fullness at the top of the tube suddenly inflates; in a flash a broad segment of snowy margin has expanded, like the opening of a hand; one more deep sigh, again that quiver through the bud, and that wide margin—gleaming white as any new fallen snow—is fully unfurled! Look along that wall of green. It is all astir; discs of snow burst forth here and there by scores; how marvelous, how human is this wakening of a flower! Who wonders that Shakespeare—that flower and nature lover—sung about "winking May buds that ope their golden eyes!"

Strange sleep of the plants, how full it is of mysteries! Sleep is called among mortals "the twin of death," but seems to be much more death's twin in the long waiting of many seeds; life lingers in them passive for so many seasons! Sometimes when locked in what is truly death, plants become again vivid in other manifestations.

There was a lily bulb once placed in the hand of a Pharaoh as he went his long journey to Rhadamanthus. After centuries of burial it was given to one of these latest summers and became a flower. There was a root of lotus held by a royal red man in his grave; his skeleton fingers yielded it to the inquisition of his white successor, and it brought forth stem, leaf, and bloom. There was a handful of wheat wrapped up in the mummy cloths of an agricultural priest, long ago. Egypt delivered it to England, and it thrived into a harvest. Surely such return of life in the least, hints of resurrection for the greater.

Turn, on a November day, to the fire in the grate. The red flames wave and curl about great lumps of blackness that were ferns, mosses, cycads many ages past. Floods of mud have buried them, earth's inner fires have charred them, yet in many the leaf impression, or the stem structure, remained as the "broad arrow" of the plant-world stamping ownership, and now and here these long dead children of the sun live again in light and heat.

We have gone surely far afield on this chilly November day! We leave the fire in the grate and go out by wood and roadside. Color comes now across the landscape, not from squadrons of parti-colored bloom, but from great red spires of sumac rising in the waste corners of the bird and plant-beloved rail fences. Oh, wretched day when wire fences took their places! The blackberry vines show brilliant strands and whips of red and purple, with clusters of dull red leaves clinging to them; the rose hips glow red as sunlit carbuncles on the swaying vines, where bronze and green leaves still linger, and here in a corner is a dandelion peeping out, and one last lingering purple aster. · Yonder Jack-in-the-pulpit holds high a thick spike of glowing coral berries, and here is the last little Benjamin of the year—the low, velvet-tufted gray balls of the " life everlasting," with a faint aromatic odor which reminds one of

the attics, presses, ponderous bureaus, and huge
cedar chests in stately ancient houses. Dr. Holmes
said this aromatic odor reminded him of the great
Pyramid and mummied Pharaohs, the asphodels of
heaven and other of the things that cannot die.

Great saucy bluejays and late-staying robins rest
and preen on gray stone walls, in the chinks of which
mosses and little ferns grow, and along which runs a
wealth of red Virginia creeper. But what of color
yet lingers on wood and pasture land, in clear blue
skies and along the edges of the ponds that reflect
them, cannot cheat us into a fancy that they will not
soon all be locked in ice or buried in snow. Nature
is making ready for her rest. We shall hear from
her lips the old answer:

" Trouble me not, for the door is now shut and my
children are with me in bed. I cannot rise and give
thee. "

CHAPTER XII

THE REIGN OF THE IMMORTALS

DECEMBER

" The wind-flower and the violet they perished long ago,
  And the briar-rose and the orchis died, amid the summer
    glow;
  But on the hill the golden-rod, and the aster in the wood,
  And the yellow sunflower by the brook, in autumn beauty
    stood,
  Till fell the frost from the clear cold heaven, as falls the
    plague on men,
  And the-brightness of their smile was gone from upland,
    glade, and glen."

WINTER and desolation are yoked together in our thoughts; snow-buried earth and leafless trees, that is our mental picture of it. Observation, really seeing what is, will greatly modify this opinion. Let us go out for a walk on a sunny day in late December. A bit of forest near the sea in southern New Jersey will be as good a place as any to study the winter beauties of the woods.

Five or six inches of snow lie on the open fields; in the forest the ground is less thickly covered.

198

There are little tracks on the soft snow, left by small animals not yet gone to winter quarters, and perhaps we find a dead fly or bee or moth, a late lingerer overtaken by the cold.

What has winter to show us?

First the lichens; these flourish as in summer time; possibly the dearth of flowers makes their green, gray, and black, with the scarlet and yellow lighting them up here and there, seem even richer and more beautiful than at other times. Long-lived and irregular in the matter of their productive periods, all seasons are alike to them.

The mosses are also as beautiful as ever; brush the snow from the dark velvet-like cushions of green, covering decayed wood, or carpeting the spaces about the thick trees. Dark green, light green, under a microscope they show many variations in their little bright sessile leaves, closely packed along the main stems. Shooting up here and there we find slim brown or reddish stalklets with a tiny cone-shaped object at the top. This is capped by something comically like an old-fashioned candle extinguisher. The wee urn is the spore case, full of green spore powder, which is kept from falling out too soon by that extinguisher-like lid. Also the spores are further protected and prevented from escaping in damp weather by a row of little teeth about the urn's top. These teeth rise and allow spores to escape in dry weather only. The moss family is varied and numerous; in temperate zones usually only the smaller members appear. Some of these are almost too minute to be seen with unaided eyes.

The beds of ground-pine are a depth of verdure; we can gather long streamers of it for our Christmas wreaths. Ground-pine, which is not a pine at all, looks like fern, and is not a fern. What is it then? A club-moss, a lycopodium. Its vivid green in the winter woods will remain upon it for weeks after it is gathered, and then it will fade to a beautiful brown with tawny or dull golden edges.

Most of the ferns have withered to the underground stems in this December weather. The winter fern, with its stiff serrated fronds, lifts stoutly above the snow; in some of the most sheltered places walking-fern, shield-fern, bracken, and lately-opened fronds of basket-fern, still linger.

The checkerberry or wintergreen, and the partridge-berry, just as deeply green in its vines and as brilliantly red in its berries, but less aromatic in flavor, have met winter as courageously as the lichens. Bright green leaves, bright red berries, smile from under the white edges of the snow-blanket.

We lift our eyes to the bare branches of the trees;

we confront a mystery. We are so accustomed to this annual fall of the leaf that " use has dulled us to its strangeness." Let us think of this work of late autumn stripping the branches. We know something of " how " it is done, but " why " is it? We are told it is a token of hard times on our earth. Once all the world was warm the year round, and all the woods were leaf-green year in and out.

Now the immortals among the trees live chiefly in the tropics. The far north of the temperate zone has a few varieties of evergreens, and our winter woods have samples of these. When the age of ice and the ages of long winters north and south of the tropics came upon the world most of the trees assumed a habit of dropping their leaves and withdrawing from business in the cold season. These bare-of-leaf trees were the glory of summer; the evergreens, the tree immortals, are the glory of winter.

Here in our winter woods that large shrub—towering into a tree betimes—the holly, attracts attention; its broad, darkly green, thorn-guarded leaves cannot hide the rich clusters of red berries. These sharp spines on the edges of the leaves are the hardened woody fibre extended beyond the green cellular filling-in.

These leaf-spines have queer tales to tell us. They

are only needed to protect the leaves from browsing
animals; the leaves will have and not have spines
in various stages of growth. The holly, when young,
has very many and stiff spines on the tender leaves,
dangerously within reach of nipping teeth; when
well grown the leaves are far less armed. The evergreen oak, as a tree, has smooth leaves. If you keep
it trimmed into a shrub it puts spines on its leaves.
A famous botanist says: "Such observations throw
us back on the unity of design in nature, leading us
away from the earth to Him who is the end of
problems and the font of certainties."

Near the holly stands that beautiful evergreen, the
juniper, abundant in hoary, blue berries. This is the
darling of our winter woods, joy of the artist's eyes,
good providence to the hungry birds. "Sweet is the
juniper, but sharp his bough;" "Azure tinted
juniper"—so sing the poets.

Around the holly and the juniper stand the tree
immortals—fir, pine, hemlock, cedar, spruce, balsam.
They have not the gay gifts of the holly and juniper,
but their brown cones, large and small, pointed or
round, compact or loosely opened, have their own
beauty.

We noted that the spines on leaves were for de-
fense, found chiefly low down, and abandoned as the
tree grows. The evergreens have another protective

method in their numerous sharp, finely-divided leaves, which we call needles. How are these protective? The tough, needle-like leaves and roughly scaled stems are uninviting to animals, but their chief value is as against snow. Early snow-storms, coming before the autumn leaves have fallen, weigh down the branches until they are broken, and often whole trees are destroyed. The evergreens of cold climates all have polished, much-divided leaves, and the snow passes between them, or falls readily from them, so that they are far less likely to be destroyed by storms than are trees with broader leaves. The evergreens of the tropics, on the other hand, usually have very broad leaves. There is a " reason why " in all things if we will but look it up.

These trees about us in the winter woods have made preparations for winter, some retaining foliage and securing it from harm, others providing for its fall without injury to the tree. The trees that shed their leaves formed at the base of the leaf-stems certain empty thin-walled cells. When the activity of the plant perished these cells quickly decayed, allowing the leaves to drop off with a clear, clean cut, while the bud for next year remained uninjured above the scar. This empty cell provision is not made by the evergreens.

As we saw in the pine, numerous leaves, bound to-

gether firmly in bundles, are well secured to the tree,
while the long needle-shape offers little resistance to
the wind. This assures their long continuance on
the tree. Do they, then, never come off? Certainly.
The pine leaves remain three or four years; the
spruce and fir five, or even seven; the yew eight;
some others are even longer lived, and remain from
sixteen to twenty years. They gradually dry and
wither for a year or two before falling, and they do
not all go in a single season. As some fall others
develop; there is continuous loss and replacement.
Thus we find the ground under the evergreens always
well-carpeted with their needles, while in undimmed
glory the green immortals among trees seem to
watch with wonder the fall of the autumn leaves
around them.

Walking among these immortals of the plant
world while the first December snow falls lightly
over them, we call to mind those "men that never
die:" Charlemagne, sitting among his peers, waiting
to return to universal empire; King Arthur at rest
in Avalon until the hour strikes for him to come
again; Genghis Khan lingers somewhere in the unknown yet to overrun Asia and Europe with his
hordes. But these are fancies only, and facts are
better. The immortals among trees give us facts of
even greater interest than these dreamings.

In Ceylon was a wonderful tree worshiped as a god, and called the Bo tree. It obtained from the natives divine honors because of its remarkable size, long life, and the fact that no other tree of its kind grew upon the island. This leads us to consider that the seed from which it sprung must have been bird-carried to Ceylon. Trees are thus found solitary, far from their kind, and owe their planting usually to birds, though sometimes on the coast their seeds were carried by the waters. The Bo tree was supposed to have been of full size and an object of worship for two thousand years. A great storm in October, 1887, overthrew it. The Ceylonese natives gathered up the fragments, cremated them, and buried the ashes with the pomp usually given to the remains of kings.

In Africa there is a tree so tenacious of vitality that nothing can kill it but fire. Livingston gives an interesting account of it. Even the severed leaves take root after they have fallen upon the ground ; stakes cut and trimmed will sprout and grow into trees if set in the ground. This tree is also known on the Island of Jamaica, and wherever it grows has a name of similar significance—the " Life Tree."

The redwoods, or sequoia, of California, are the stateliest immortals of the Western tree-world. They are cone-bearing trees, related to the pines, and standing between the firs and cypresses. The California redwood, one of the two varieties of sequoia, grows to gigantic size ; a trunk is recorded two hundred and seventy feet high and fifteen feet in diameter, and even this enormous height and girth are said to have been surpassed.

The redwood bark gives the tree its common name, being red, like the Scotch fir, deeply ridged and twisted, as is to be expected of the scarred and wrinkled veteran of so many ages. When the sequoias are young they are very graceful and beautiful, bearing their branches in a regular cone shape, the lower ones sweeping the ground ; the leaves are flat, linear, and very glossy, while from under the twigs hang the cones, nearly two inches long, bluish-green when young, and of a rich seal-brown in maturity. The catkins are on the tips of the twigs, appear in June, are round, and of a light-brown shade.

The botanical name of this tree, sequoia sempervirens, refers to its perennial green. It flourishes best on the California coast-line. Long-lived and majestic as this tree is, its cousin, the other sequoia, known as the gigantea, surpasses it in size. This is a native of the mountains, growing on the slopes of the Sierra Nevada range. The gigantea is a social tree, seldom found solitary ; it is best shown in

groves, on a plateau about five thousand feet above
sea-level. The needles or leaves are smaller than
those of the sempervirens, and are more rigid; the
catkins are also small. Growing at so great a height
snow-storms are no stranger to it, and the stiff,
bristle-like form of the needles is best fitted to
encounter the driving snows. The gigantea often
attains a height of four hundred feet with a diameter of thirty. Sequoia-wood is strong and dura-
ble, of a rich color, especially the heart-wood or
middle of the trunk; it is capable of receiving a
rich, glossy polish, and thus is a favorite for cabinet
work. One of these trees, called by the Indians
"The Mother of the Forest," measured ninety feet
in girth and was three hundred and twenty-one feet
high. The bark was stripped off for a space of one
hundred and sixteen feet to be shown at an exhibi-
tion, and thus this stateliest tree of the American
forests died. The growth of the redwoods is rapid
for about twenty years; after that growth is more
and more slow, and in age the increase is very tardy.
Age also robs sequoias of their grace and beauty;
the lower branches fall away, and the foliage becomes
dryer and more sparse. The life of a redwood tree
is estimated at about three thousand years, if left
unharmed by men and forest fires.

These redwoods have rivals. Cowthorpe, in

England, boasts an oak tree fifteen hundred years old and seventy-eight feet in girth. Under its stately branches that philosophic and Christian gentleman, John Evelyn, used to muse on the follies of his times, and the singular aberrations of the later Stuart kings. South America also claims one of these immortal giants, a tree one hundred and twelve feet in girth, and said to be four thousand years old; but trees are like centenarian people—when they pass the ordinary line of life, a few years may be recklessly added to their age, and no one can, with certainty, offer contradiction! Even such plants as are usually considered short-lived may, in especial instances, reach and pass their century—as grape, rose, ivy vines, and thorn-trees are known that have lived their two centuries; but the thorn-rose and grape shed their leaves and look dead enough for several months in the year. Only the glorious evergreen plants deserve to be ranked in the cohorts of the immortals.

THE END

www.ingramcontent.com/pod-product-compliance
Lightning Source LLC
Chambersburg PA
CBHW021942220326
41599CB00013BA/1486